COURSE

IN

ELEMENTARY PHYSICS.

BY

CHARLES R. CROSS,

Assistant Professor of Physics in the Massachusetts Institute of Technology.

BOSTON:

PRESS OF A. A. KINGMAN.

1873.

COURSE

IN

ELEMENTARY PHYSICS.

BY

CHARLES R. CROSS,

Assistant Professor of Physics in the Massachusetts Institute of Technology.

BOSTON:
PRESS OF A. A. KINGMAN.
1873.

PREFACE.

The present pamphlet comprises the text of the first few chapters of an elementary work on Physics, designed to supplement the lectures delivered to the students of the Massachusetts Institute of Technology during the first two years of their course. It is intended to impart such a general knowledge of the science as will enable them with advantage to continue the pursuit of the subject in the more advanced work of the Physical Laboratory.

It will be the aim of the author to present the student with a concise statement of the leading principles of the science, and to explain the methods of investigation by which they have been ascertained.

In the preparation of the work constant reference has been made to the original memoirs as well as to the standard text books upon Physics. In the chapters relating to mechanics, the *Elements of Mechanical Philosophy*, by Prof. W. B. Rogers, long out of print, has been of the greatest service, and valuable hints have been derived from the elementary works of Smith, Peck and Mayer. The text-books of Thomson and Tait, Rankine, Todhunter and Price have furnished many suggestions.

A list of standard text-books upon General Physics will be added hereafter, as well as a collection of physical tables. The references appended to various chapters are not intended to be exhaustive, but may serve as aids to the student who is desirous of investigating the subject more thoroughly.

The work will be put in a more permanent form, and properly illustrated, as soon it reaches a suitable stage of progress. The present sheets have been printed for the temporary use of the students.

Boston, January, 1873.

ELEMENTARY PHYSICS.

CHAPTER I.

General Methods of Physical Science.

1. Physical Sciences. All those branches of human knowledge which have for their end the study of material objects and the phenomena of the external world, are included among the Physical Sciences. They may be divided into two general classes; first, those which aim simply at a classification of objects, and, secondly, those which endeavor to ascertain the laws of the phenomena of the material universe. The former class comprehends the various branches of Natural History. It includes Zoology, which describes and classifies the various forms of animal life; Systematic Botany, dealing with the classification of plants; Anatomy, both vegetable and animal, describing the structure of the organs of living things; and Mineralogy, which concerns itself with the constitution of the rock masses of the globe. The second class comprises Mechanical Philosophy or Physics proper (of which Astronomy is a branch) and Chemistry, which investigate the general laws of natural actions; Dynamical Geology, which studies the laws regulating the structure of the earth, and Physiology, whose province is the determination of the functions and mode of action of the organs of animals and plants.

2. Physics. In its original and most extended sense, Physics, or Natural Philosophy, as it is sometimes called, comprehended the entire range of physical as distinguished from mental science. The term is still occasionally used in a comprehensive manner to denote the second class of sciences of which we have spoken.

In its limited sense, and the one in which it is used in this work, it is applied only to that branch of science which investigates the laws of physical actions in which there is no permanent change in the general properties of the body acted upon. It is thus distinguished from Chemistry, which deals with actions in which the properties of the body affected are permanently modified or entirely changed. A few illustrations will suffice to make this plain. An unsupported body falls to the earth. Here we have a physical action in the motion produced, but the general properties of the body are not altered thereby, for it has the same weight, color and other characteristics as before. The law of its downward motion is then a subject lying within the province of physics. If the body is elastic it rebounds on reaching the earth; there is still no change of its general properties, so that the laws of elasticity are physical and not chemical. The motion ensuing when a piece of iron is brought near to a magnet, and the vibrations of a musical string, are additional examples. On the other hand, if a spark be applied to a mixture of oxygen and hydrogen they unite with an explosion, forming an entirely new substance, water, possessing properties totally different from those of either of the gases composing it. This, then, is a chemical change. Again, iron when exposed to moist air rusts, because of the union of oxygen with the metal; a new body, oxide of iron, is thus obtained, unlike either constituent. The laws of such changes are chemical laws. As would naturally be expected, there are many cases in which it is difficult to decide whether a given action is chemical or physical, or a mixture of both. We also have occasion to examine the physical effects of chemical changes, so that there are lines of investigation in which it is difficult to decide where one ends and the other begins.

3. Deductive and Inductive Sciences. The various sciences may be divided into two general classes, according to the mode of reasoning employed in the investigation of the facts pertaining to them. These methods are known as *deductive* and *inductive reasoning*, and the sciences characterized by their respective use as the Deductive and the Inductive Sciences. Deductive reasoning starts with certain general fundamental facts and definitions, and from these evolves truths which follow necessarily from the premises. Induction observes particular facts, and by analyzing and comparing them rises to more general truths. Deduction reasons from the general to the particular; induction from the particular to the general. Pure Mathematics furnishes us with an example of a deductive science. Certain definitions are laid down, and from these in connection with the axioms, or fundamental truths relating to magnitude, the whole science is evolved without the necessity of going outside of the mind itself for any portion of the demonstration. The most abstruse theorems follow directly from the fundamental simple ones. The Physical Sciences, on the contrary, are based on induction. Their end is to form a natural classification of objects, and to ascertain the general laws of the forces acting upon matter, and in no way can this be done except by the comparison of numerous particular cases. Hence we must reason from these to general laws. For example, a natural classification of animals can be attained only by observing the different species, noting their fundamental points of agreement and difference, and thus arranging them in groups connected with each other by certain structural peculiarities, which at the same time separate them from all other animals.

Having thus briefly stated the difference between these two kinds of reasoning, let us consider more fully the process of investigation by which the general laws of the physical world are ascertained and logically arranged.

4. Physical Law, Theory, Hypothesis. We shall frequently have occasion to use the terms *law*, *theory* and *hypothesis*. By a physical law is meant simply the constant method according to which a cause acts. "It is the custom of philosophers whenever they can trace regularity of any kind to call the general proposition which expresses the nature of that regularity a law, as when in mathematics we speak of the law of decrease of the successive terms of a converging series."[1] A theory is a true, complete and philosophical explanation of connected phenomena and their laws; as, for example, the theory of gravitation or the theory of sound. A hypothesis is a supposition as to the nature of any law or cause made in the absence of absolute proof, as, for example, the hypothesis of an electric fluid. When a hypothesis is extended to include a number of phenomena, and is proved to be true, it becomes a theory. The term theory is frequently applied more loosely than in the above definition to denote any plausible and clearly defined hypothesis.

5. Observation and Experiment. The facts which form the basis of physical science are ascertained in two ways, by *observation* and by *experiment.* Observation is the careful examination of phenomena as they occur in the order of nature; experiment is the production of phenomena by causes under our own control. It is evident that the latter method of investigation is a great extension of the former, and will often enable us to ascertain laws which could never be known without its aid. To illustrate by an example, suppose that we wish to know which of the components of the atmosphere, oxygen or nitrogen, enables it to support animal life. To ascertain this two processes are open. We may (1) notice which constituent of the atmosphere disappears when an

[1] This quotation, together with several others in the present chapter, is from Mill's *Inductive Logic.*

animal is immersed in a limited quantity of air, or (2) we may immerse animals in oxygen and in nitrogen and compare the effects. Both of these methods require us to perform an experiment, and as nature does not furnish either oxygen or nitrogen in a pure state, and also as the oxygen consumed by animals from the atmosphere is replaced by that given out by plants, we could never have ascertained the truth in question from pure observation. Indeed, without the process of experimentation we could not even know of the existence of two ingredients in air. In some sciences, however, as Astronomy, the former method is the only one possible.

6. Induction. Having by these methods examined a large number of phenomena, we next approach the process by which we determine the causes of these phenomena and their laws of action. The great logical principles upon which our procedure is based are (1) that "every fact which has a beginning has a cause," and (2) that "the course of nature is uniform," the first of these being an intuitive idea, the second a truth ascertained from universal experience. Hence, when under certain influencing circumstances a given event occurs, we are entitled to infer that it will recur under the same circumstances, and it is the first object of science to ascertain exactly what these circumstances are, for they are the *causes* of the phenomena. In physical science we limit ourselves strictly to what are known as *physical causes,* that is the phenomena which are invariably and necessarily followed by some other phenomenon. The criterion by which we know that any combination of phenomena is the cause of another phenomenon is, that between the first and last there is invariable and *unconditional* sequence. It is not sufficient that the sequence be invariable, for in this case both events might be consequents of a single cause. Thus day and night invariably follow each other, but do not stand in the relation of cause and effect. The antecedence of the supposed cause must also be unconditioned, that is, its relation to the effect must be demonstrably such that without it the effect could not be produced. This demonstration can be given only by experiment or by the process of deduction. The constant endeavor of science is to find the smallest number of causes that will account for all natural phenomena, or as Mill expresses it, to find "the fewest and simplest assumptions which, being taken for granted, the whole existing order of nature would result." These primary assumptions we call Laws of Nature.

The causes of the phenomena studied by the processes of observation and experiment are to be ascertained only by a rigorous analysis and comparison of the circumstances attending their production. By investigating a number of isolated cases and ascertaining those circumstances which are present in all, we obtain a more general result, which we may infer to be true in analogous instances, even though these may be beyond the range of our present observation.

A single example will illustrate this. Experimental tests show that when oxygen and hydrogen unite to form water, the proportions are invariably 8 parts by weight of oxygen to 1 of hydrogen. So also when chlorine and hydrogen unite to form chlorhydric acid, the invariable proportion of the constituents is 35.5 parts of the former to 1 of the latter. In all other chemical compounds which we have examined, a like constancy of proportion is observed. Hence we lay it down as a general law, that the proportions in which substnaces unite to form a chemical compound are fixed, definite and invariable; and are justified in predicting that the law will hold for any unknown chemical compound that may be discovered in the future.

There are several processes too often confounded with induction which should be carefully distinguished from it. (1.) Any logical operation which contains nothing in its conclusion which is not stated in the premises from which it is drawn, *i. e.*, in which the conclusion is not wider than the premises, is not properly induction. Thus were we simply to say that all chemical compounds *which we have examined* combine in definite proportions, this would not be an induction. It would be merely a short way of expressing the several facts already known, that oxygen and hydrogen, etc., combine thus. We perform a process of inductive reasoning only when we infer from the known instances to all instances of chemical combination. (2.) Mathematical proofs, though leading to general propositions, are not inductions, because there is no inference from known to unknown cases. Suppose, for example, that we have proved that the three angles of a triangle are equal to two right angles. This is in reality only shown to be true of thé particular triangle before us in our mind, or drawn on paper, but since we see that the same method of proof will apply to any other triangle we assume the theorem to be a general truth. The reasoning is not, "All the triangles that I know have the sum of their angles equal to two right angles, hence I infer that all triangles possess this property," but, "The method of reasoning applied to this particular triangle evidently applies to all possible triangles, and hence the theorem is true of all." Mill suggests that if this is to be allowed the title of induction at all, it should be called *Induction by parity of reasoning*. (3.) Another process often confounded with induction is that which is called by Whewell the *Colligation of Facts*, which is a mere description of observed facts, and not an induction from them. To illustrate, when the astronomer Kepler announced as the result of his observations on the planet Mars that it moves in an ellipse, he simply expressed in that one statement that the successive positions of the planet in the heavens lie in such a curve. There was no inference of unknown from known. Instead of saying that on successive evenings Mars occupied certain definite positions, and stating these in order, he said that these positions are such as would be occupied by a body revolving about the sun in an ellipse of known dimensions and position.

7. Deduction. The general laws of any series of phenomena once obtained by the inductive process, we are frequently enabled to extend our knowledge very greatly by the application of deductive reasoning. For assuming the truth of the law in general, we may calculate what effects should occur under given conditions of its action, and confirm these calculations by direct experiment. Thus, to take a very simple instance, supposing the chemical law of definite proportions to be proved, on finding any new substance we may assume that if its constituents are chemically united, their proportions are the same in all specimens, and if on analyzing different samples of it we find this is not the case, we may justly conclude that the union is not purely chemical, thus obtaining a new fact regarding the substance.

If the results of deduction coincide with those of observation and experiment, the strongest confirmation of the theory in question is produced. In case of a want of coincidence, it does not necessarily follow that the theory is false, inasmuch as there may be certain circumstances which we have neglected to take into account in our calculations.

Deduction often brings facts to our knowledge which it would be very difficult, if not impossible, to ascertain by inductive reasoning alone, since it directs us to the proper course of experimentation in any given case. In the "interrogation of nature" everything depends upon the question put, as the only answer given is, "Yes" or "No." Deduction informs us as to the question which we should ask. As an example in illustration, suppose that we wish to know the maximum amount of work which could be obtained with a perfect steam-engine burning a known quantity of fuel. We are incapable of answering this question by direct experiment until we can construct a perfect engine, and even in that case it would be impossible to prove it to be perfect by inductive methods. If, however, we accept the truth of the law, that a given quantity of fuel when burnt can produce only a definite quantity of mechanical work, we have an immediate answer, since a perfect engine would utilize all the heat produced by the combustion, and the equivalent of this heat expressed in units of mechanical work would evidently be the amount required. Our experimentation should therefore be directed to the obtaining of this equivalent.

From what we have said it will be seen that the process of deduction to be complete must embrace three distinct operations; (1) Induction, (2) Ratiocination, (3) Verification. As the basis there must be some law or laws previously ascertained by induction as general facts from which the unknown particulars may be deduced. Hence to pursue such a deductive investigation we have first to learn the law of each separate cause which contributes toward the unknown effect.[1] Next comes the determination of the effects from a knowledge of the causes, which is an operation of reasoning often of the utmost complication, and which even in some of those cases apparently the most simple, is so baffling that we have as yet obtained only approximate results. Finally, every result of deductive reasoning should be verified by observation or experiment, that we may know whether in our calculations we have neglected to consider any of the attendant circumstances.

[1] The case of deduction from hypothesis is considered in § 9.

8. Application of Mathematics. The methods of deduction assume an incalculable importance in Physics, because we are thus enabled to apply the processes of mathematics to determine what will be the result of the action of various concurring forces whose laws have already been ascertained. The results thus obtained will be equally certain with the prior inductions on which the general laws are based; provided, of course, that no elements are omitted from the calculation. It should, however, be observed that the remark in the preceding paragraph concerning approximate solutions frequently applies to this use of mixed mathematics, since even the most subtle processes of the calculus often fail to furnish more than these. Those subjects to which mathematics is thus applied, are frequently known as the *Physico-mathematical Sciences*, and there are now few portions of Physics which have not a more or less developed mathematical theory.

9. Hypothesis. In the absence of certainty as to the causes or mode of production of phenomena, we frequently make use of hypotheses regarding them, and then by the application of deductive processes ascertain whether the result of these calculations coincides with the facts obtained by observation. That is, in the course of reasoning explained in § 7, we substitute a supposition for the first process there mentioned, that of a prior induction. If the supposition with which we start is reasonable, this becomes a valuable addition to our methods for the discovery of truth. But it should be borne in mind that the mere fact of coincidence between observed phenomena and the results deduced from the hypothesis does not necessarily show that the assumed hypothesis is true. To prove this it must also be demonstrated that no other hypothesis will explain the known facts, and unless the supposition is of a nature to allow of such a demonstration, it is not perfectly satisfactory, since it is capable neither of being proved nor disproved. Any hypothesis, however, if plausible, may be of great service as a guide to experimentation.

10. Explanation of Laws. In speaking of physical phenomena we constantly make use of the terms *explanation of phenomena* or *explanation of laws*. It is desirable that a clear conception should be formed of the meaning of these. Any fact is said to be *explained* when the laws of causation on which its production depends are stated. In like manner, a law of nature is explained by showing it to be a particular case of a more general law. There are three methods of explanation of laws which apply in different cases. (1.) The first is that in which several causes unite to produce a joint effect equal to the sum of the effects which each would produce were it acting alone. In this case "the law of complex effect is explained by being resolved into the separate laws of the causes which contribute to it." Thus in the consideration of curvilinear motion we shall explain the law of the movement of a planet about the sun by showing the motion to be the result of the joint action of two forces, one of which is continually drawing the planet towards the sun, while the other continually tends to cause it to fly off in a tangent to its orbit. The orbital motion is thus *explained*, as it is shown to result from forces with whose laws we are already familiar. (2.) "A second case is when between what seemed to be the cause, and what was supposed to be its effect, further observation detects an intermediate link; a fact caused by the antecedent, and in its turn causing the consequent, so that the cause at first assigned is but the remote cause, acting through the intermediate phenomena." The bleaching action of chlorine is explained in this manner. Between that gas and certain bases, specially hydrogen, there is an affinity so great, that chlorine will decompose most compounds of the metallic bases or hydrogen. These are constituent elements of almost all coloring matters, so that the latter are decomposed by chlorine, and their color caused to disappear. Here, the intermediate fact is the affinity of chlorine for hydrogen. In like manner the efficacy of chlorine as a disinfectant is explained, since hydrogen is an essential component of the infectious matter. (3.) The third method of explanation is what has been called the *subsumption* of one law under another. It is "the gathering up of several laws into one more general law which includes them all." As the best possible example of this we cite the generalization of Newton, proving that the force acting to keep the planets in their orbits by tending to draw them towards the sun, and that causing a body to fall to the earth, are both due to an attraction exercised by every particle of matter in the universe upon every other particle.

If the student on reaching that point will carefully read the section of this work treating of the history of the law of universal gravitation, in connection with the present chapter, he will find excellent examples of almost every process which we have described in the preceding pages, as the series of inductions which led to the establishment of that law is the most brilliant in the history of science.

Our explanations, as we call them, give us no insight into the inner causes of phenomena. "What is called explaining one law of nature by another, is but substituting one mystery for another, and does nothing to render the general course of nature other than mysterious; we can no more assign a *why* for the more extensive laws, than for the partial ones. The explanation may substitute a mystery which has become familiar, and has grown to *seem* not mysterious, for one which is still strange. And this is the meaning of explanation in common parlance. But the process with which we are here concerned often does the very contrary; it resolves a phenomenon with which we are familiar, into one of which we previously knew little or nothing, as when the common fact of the fall of heavy bodies is resolved into a tendency of all particles of matter towards one another. It must be kept constantly in view, therefore, that when philosophers speak of explaining any of the phenomena of nature, they always mean (or should mean) pointing out not some more familiar, but merely some more general phenomenon of which it is a partial exemplification, or some laws of causation which produce it by their joint or successive action, and from which, therefore, its conditions may be determined deductively. Every such operation brings us a step nearer towards answering the question, which was stated some time ago as comprehending the whole problem of the investigation of nature, *viz.*: What are the fewest assumptions which being granted, the order of nature as it exists would be the result? What are the fewest general propositions from which all the uniformities existing in nature could be deduced?

"The laws, thus explained or resolved, are sometimes said to be *accounted for;* but the expression is incorrect if taken to mean anything more than what has been already stated. In minds not habituated to accurate thinking, there is often a confused notion that the general laws are the *causes* of the partial ones; that the law of general gravitation, for example, causes the phenomenon of the fall of bodies to the earth. But to assert this would be a misuse of the word cause; terrestrial gravity is not an effect of general gravitation, but a *case* of it; that is, one of the particular instances in which that general law obtains.[1] To account for a law of nature means, and can mean, no more than to assign other laws more general, together with collocations, which laws and collocations being supposed, the partial law follows without any additional supposition."[2] The cause of these ultimate laws of action is to be found only in the design of the ruling Mind of the universe.

11. Construction and Correction of Theories. In the mathematical discussion of physical questions, we generally construct our theories by first considering only the simpler and more essential elementary forces acting to produce the phenomena in question. These discussed, other forces which merely modify the general result are taken into account in the form of corrections. To illustrate, suppose we wish to deduce the curve followed by a body, as a cannon-ball, projected into the air, a problem of the greatest importance in the science of gunnery. The consideration of the two principal forces acting upon the ball, viz., the original force of projection, tending to make it proceed in a straight line, and the attraction of the earth, which constantly draws it towards the ground, shows that the path would be a certain mathematical curve known as the *parabola*, were there no disturbing forces.

Such disturbing forces exist, however, in the resisting air through which the projectile moves, so that it is only in a vacuum that the parabolic curve could be strictly followed. The effect of atmospheric resistance in altering the path must then be taken into account as a correction to the original theory, which can be done if we know the relation between the retardation and the velocity of the projectile. Often, as in the case under consideration, these corrections are very difficult to apply, owing to our ignorance of the exact law of the disturbing action, or to complexities in its application when known. In like manner, when considering the theory of machines, we first suppose their component parts to possess no weight, to be absolutely rigid or perfectly flexible, and to act upon each other without friction, and we determine the law of the transmission of power under these conditions. We afterwards ascertain what loss of power there will be on account of friction, and also estimate the effect of the weight of the different parts, and their want of perfect rigidity or flexibility. Not until all these

1 It is customary, however, to speak of the fall of bodies as being caused by *gravitation*, meaning by this not the law, which is merely the statement of the general fact, but that force (of the nature of which we are profoundly ignorant) which causes the tendency of particles of matter to approach each other.

2 Mill, *System of Logic*, Vol. I., Bk. III., p. 529, 6th Ed.

corrections have been applied to the original theory, does it meet the requirements of practical science.

12. Discussion of Observations: Analytical and Graphical Methods. After making a series of experiments or observations with the intention of determining the law of any phenomena, there are two methods of mathematical discussion that we may apply to them. These are (1) the *Analytical*, and (2) the *Graphical Method.*

Analytical Method. In applying this method, we make use of algebraic symbols, designating each one of the quantities considered by a letter, and discussing their relations by the analytical processes of pure algebra or the calculus.

Graphical Method. This is geometrical, the quantities considered being represented by lines, from whose relations the relation of the quantities themselves is deduced.

The first of these methods is more accurate than the second, but is at the same time more difficult of application owing to accidental errors in observation, and the complexity of many laws. The graphical construction appeals at once to the eye, and is frequently of the greatest service, because of the ease with which the relations among the lines of the figure, and hence among the quantities, are determined.

As a simple illustration of the use of the graphical method let us take the case of a daily record of thermometrical observations extending through a month. We draw two lines, OX and OY, Fig. 1, at right angles to each other, for purposes of reference. These are known as the *axis of* X and the *axis of* Y, respectively. Their point of intersection O is called the *origin of coördinates*. Lines drawn from OY parallel to OX are called *abscissas*, which are positive to the right, and negative to the left of OY, and lines from OX parallel to OY, *ordinates*, which are positive above, and negative below OX. We represent times by abscissas and temperatures by ordinates, laying off on OX spaces proportional to the time from the beginning of the observations, and parallel to OY spaces proportional to the temperatures at those times. Passing a curve through the points thus determined, we have the desired *thermogram*, or temperature curve. The curve of Fig. 1, is constructed from the data given in the annexed table, and represents the mean daily temperature on the summit of Mt. Washington during the month of January, 1871.

Date.	*Temperature.*	*Date.*	*Temperature.*	*Date.*	*Temperature.*
Jan. 1.	—3.0	Jan. 11.	21.5	Jan. 21.	10.5
" 2.	8.2	" 12.	32.5	" 22.	—28.5
" 3.	—1.6	" 13.	36.0	" 23.	—2.6
" 4.	—14.8	" 14.	33.5	" 24.	—2.5
" 5.	8.7	" 15.	30.2	" 25.	—15.9
" 6.	12.5	" 16.	31.7	" 26.	3.8
" 7.	—7.2	" 17.	2.7	" 27.	—4.5
" 8.	—5	" 18.	1.5	" 28.	—5.6
" 9.	—7.7	" 19.	8.0	" 29.	6.5
" 10.	0.7	" 20.	15.5	" 30.	10.5
				" 31.	27.7

It is convenient to record many continuous phenomena in this manner.

As an application of the graphical method to the discussion of physical laws, we give the following illustration. Let it be required to find the relation between the volume of a quantity of gas and the pressure to which it is subjected. Suppose that experiments have given the results shown in the following table.

Pressure.	*Volume.*	*Pressure.*	*Volume.*	*Pressure.*	*Volume.*	*Pressure.*	*Volume.*
30	100.00	50	60.00	70	42.86	100	30.00
32	93.75	55	54.34	80	37 50	110	27.27
36	83.33	60	50.00	90	33.33	120	25.00
40	75.00	65	46.54				

Denote pressures by abscissas and volumes by ordinates, and construct the curve AC, Fig. 2. This will be found to be an *equilateral hyperbola*, of which OX, OY are asymptotes, a curve having the property that the product of the ordinate at any point by the corresponding abscissa, is a constant. Since the ordinates have been made proportional to the volumes of the gas, and the abscissas proportional to the pressures, it follows that when a gas is compressed, the product of its volume by the pressure upon it is a constant, which is the law sought.

A consideration of Fig. 2 will also show an easy manner of finding the volume of gas corresponding to any pressure not given in the table. Thus if we wish to ascertain the volume under a pressure of 45, we simply lay off $Oa = 45$, and draw ab perpendicular to OX till it meets the curve at b: the line $ab = 66.7$, represents the volume wished, since $Oa \times ab$ is a constant, and Oa has been made proportional to the given pressure. From this example it will be seen that the graphical method gives a simple way of determining intermediate values of a variable, or *interpolating*, as it is called.

List of Works of Reference.

For further information concerning the subjects treated in the preceding pages, the student may consult,

A System of Logic, by John Stuart Mill; Book III., *On Induction.*

Preliminary Discourse on the Study of Natural Philosophy, by Sir J. F. W. Herschel.

Philosophy of the Inductive Sciences, by Wm. Whewell.

History of the Inductive Sciences, by Wm. Whewell.

Cours de Philosophie Positive, par Auguste Comte, *Leçons* 1–39. (*Astronomie, Physique, Chimie.*)

More extended notices of the Analytical and Graphical Methods, and the general discussion of observations will be found in

Lecture Notes on Physics, by Prof. A. M. Mayer (Phila., 1868), pp. 28–49.

Elements of Physical Manipulation, by Prof. E. C. Pickering (Boston, A. A. Kingman, 1872), pp. 1–25.

In these two works a number of valuable references are given to articles bearing upon the subject in question.

CHAPTER II.

Weights and Measures.—Instruments of Measurement.

13. In physical investigations it is constantly necessary to determine various magnitudes, so that it will now be advisable to consider the two systems of weights and measures in common use, which are known as the English and the French System.

English System; Linear Measure. The unit of English linear measure is the *yard* of 36 inches, a length which has come down to us from very early times.

The statutory magnitude of the yard is said to have been originally determined in 1120 by the length of the arm of Henry I. There being a great discrepancy among the different yards in use in Great Britain, Graham, in 1742, after a comparison of all the yards and ells of best authority, constructed a standard of which Bird made two exact copies, one in 1758, and a second in 1760. An Act of Parliament in 1824, declared that the length of the yard constructed in 1760, when at a temperature of 62° Fahrenheit, should be considered as the legal standard of linear measure for the kingdom, providing for its recovery in case of loss by making it to consist of 36 inches of *such length* that 39.13929 of them should be equal to the invariable length of the pendulum beating seconds at London, in a vacuum, and at the level of the sea. This Act therefore made the length of the seconds pendulum the basis of the English System. The length of such a pendulum had been carefully measured by Capt. Kater, who determined its value in standard inches of the scale of 1760 as given above. The standard scale, together with that of 1758, was destroyed by fire on the burning of the House of Parliament in 1834, and it became necessary to construct a new one. The scientific committee to whom the matter was referred, did not deem it expedient to adopt the method of restoration by the use of the pendulum, as provided in the Act of 1824, because serious sources of error had been shown to exist in the processes used by Kater in his measurements, rendering it impossible to verify these within $\frac{1}{500}$ of an inch, while a copy of any existing scale could be constructed so as to vary far less than this amount from the original. They therefore advised that all the most authentic copies of the former standard should be compared, and a new one constructed from these, which was accordingly done. They also advised that the new standard should in no way be defined by reference to any natural basis, such as the pendulum, and in the Act of 1855 legalizing the new standard, the yard is defined as being the distance at 62° Fahr., between two marks on a certain standard bar. The British yard, therefore, is no longer based on the length of the seconds pendulum at London, but is simply a copy of a material scale kept in the Exchequer Office at Westminster.

The unit of linear measure adopted by the United States is essentially identical with that of Great Britain. The standard at the Office of Weights and Measures at Washington is a brass scale 82 inches in length, constructed by Troughton for the Coast Survey.

The distance from the 27th to the 63d division of this at 62° Fahr., is equal to one U. S. yard. Two copies of the new British standard, one of bronze and one of iron, presented to the United States by the English government, compared with the Troughton scale, show that the U. S. standard exceeds the present British one by 0.00087 in.

The units for the measure of surface and solidity, as the square and cubic foot or yard, are obtained directly from the linear unit.

14. Weight. The unit of weight at present adopted by Great Britain is the *Avoirdupois pound* of 7000 grains, the standard being a platinum weight kept in the Office of the Exchequer.

The oldest English legal standard known was the *Tower pound,* as it was called, which was used even before the Norman Conquest. and was somewhat lighter than the present *Troy pound* of 5760 grains. It was employed at the Mint as the standard for coinage up to the time of Henry VIII., by whom it was abolished in 1527, though the Troy pound was more or less in use, at least as early as the reign of Henry V. (1413–1422). There was also at this time another pound in general use, called the *libra mercatoria,* weighing about 6750 grains. When the Avoirdupois pound was first introduced is not definitely known, but it appears in statutes of the time of Edward III. (1327–1377), and like the Troy pound was probably introduced by the Lombard bankers of that day. It seems to have derived originally from a Greek weight, the *mina,* of 6945.3 grains, which the Romans subdivided into 16 ounces, while the Troy pound came from the Roman weight of 5759.2 grains, the $\frac{1}{125}$ part of the *Talent of Alexandria,* which they divided into 12 ounces. During the reign of Elizabeth (in 1582 and 1588) Avoirdupois weights were deposited in the Exchequer, but the Troy pound continued to be the legal standard.

All the early weights were carefully examined by Graham in 1742, and in 1758 a Parliamentary Committee recommended the construction of a new standard Troy pound, by which all other weights should be regulated. The bill carrying these recommendations into full effect was never passed, but three one-pound weights were made, one of which was deposited in the Parliament House. In 1818. the question was revived, and a committee, among whose members were Young, Wollaston and Kater, was appointed to investigate the subject. They advised that the Troy pound of 1758 should be the standard for the kingdom, and that the Avoirdupois pound, which had never been legally defined, should be declared to contain 7000 grains Troy. They also defined the grain weight by adopting the results of Shuckburgh and Kater, according to which a cubic inch of distilled water, weighed in air by brass weights at a temperature of 62° Fahr., with the barometer at 30 inches, was equal to 252.458 grains, of which 5760 constituted a Troy pound. In case of loss, the standard pound was to be restored by reference to the weight of a cubic inch of water. Their report became a law in 1824, but only ten years later the standard pound of 1758 was destroyed by fire. In 1838, Baily, Sir J. F. W. Herschel and others were requested to take into consideration some means for the restoration of the lost standards. In 1841, they reported adversely to restoring the pound in the manner designated in the Act of 1824, owing to the great discrepancy existing among various determinations of the weight of the cubic inch of water, which rendered the accuracy of Shuckburgh's results very questionable. In fact, the weights assigned to a cubic inch of distilled water by different experimenters exhibit a variation amounting to $\frac{1}{1200}$ of the total amount, while a copy of any existing standard could be made which should not vary from the original by more than $\frac{1}{1000000}$ of its actual weight. The committee therefore advised that the new standard be made by a comparison of the best authentic copies of the old standard. They also recommended that the Avoirdupois pound be made the legal standard rather than the Troy pound, since the former was in much more general use. The construction of the new standard was undertaken in 1843 by Prof. W. H. Miller, who collected thirteen authentic copies of the original Troy pound of 1758. Eleven of these were finally rejected from consideration, as they were made of brass and were considerably oxydized, and the two remaining, which were of platinum, were the only ones used in the determination of the exact weight of the standard of 1758. From measurements of these, a Troy pound of 5760 grains was constructed of platinum, which was regarded as identical with the former standard, and from this a platinum standard equal to $\frac{7000}{5760}$ of the weight of the Troy pound was made, together with four copies. This was constituted the legal standard Avoirdupois pound of Great Britain by the *Weight and Measure Act* of 1855. Experiments made upon the actual standard when completed, showed it to really weigh 7000.00093 grains, of which the old standard contained 5760.

The United States Standard is the Troy pound of the U. S. Mint, copied by Capt. Kater from the British Troy pound of 1758. An Act of Congress of 1834, defines it as being equal in weight to 22.794422 cubic inches of distilled water at its maximum density, but the incorrectness of the weight of a cubic inch of water under those circumstances, which was assumed by that act, causes the definition of the actual pound weight to be in error.

15. Capacity Measures. The British unit for all measures of capacity is the *Imperial gallon*, which contains 10 Avoirdupois pounds of distilled water at a temperature of 62° Fahr., with a barometric height of 30 inches. According to the best determinations, it contains 277.123 cu. in. under those circumstances. The Act of 1824 stated its contents as 277.274 cubic inches, which has since been shown to be an incorrect value. The U. S. Standard is the *wine gallon* for liquids, and the *Winchester bushel* for dry measure, derived from the English measures of the same name.

The former measure is defined by Act of Congress as containing 231 cubic inches, the latter 2150.42 cubic inches at 39° Fahr., 30 inches barometer. But unfortunately the capacity of the *British* Winchester bushel is 2150.4 cubic inches at 62° Fahr., so that at our standard temperature it contains only 2148.9 cubic inches, and at no common temperature are the two bushels equal. The same is the case with the wine gallon. The British Imperial bushel, a measure in common use, but not a legal standard, contains 8 Imperial gallons or 2216.984 cubic inches at 62° Fahr.

The method of ascertaining the number of cubic inches contained in the standard gallon or bushel, is by finding the weight of the quantity of distilled water necessary to fill it under given circumstances of temperature and atmospheric pressure. This weight divided by the weight of a cubic inch of water, gives the capacity. This method evidently requires a precise knowledge of the weight of a cubic inch of water, which, as we have stated, is so difficult of exact determination that somewhat different results have been obtained by the most skilful experimenters, causing a slight variation of the actual from the legal contents of the U. S. standards. The greatest possible deviation of these from the statutory capacity is, however, so small as not to be of very great practical consequence, the gallon and bushel never being used in any but commercial measurements.

There has been even more variation in the values of the various capacity measures of past times than among weights. The capacity of the ordinary wine gallon has gradually risen from 216 cubic inches, its contents in 1299, to its present value of 231 cubic inches. The Winchester bushel has had equally variable values, having risen from 2114.68 cubic inches in 1266, to 2150.4 cubic inches, its present capacity. The British Imperial standard gallon is a unique measure, and is derived directly from the Avoirdupois pound, while the other measures were originally obtained from the Tower and Troy pounds.

16. Circular or Angular Measure. The unit of angular measure is the *degree*, which is the $\frac{1}{360}$ of a circumference. It is subdivided into 60 *minutes*, and each of these minutes into 60 *seconds*, smaller angles being indicated by decimals of a second.

Until within two or three centuries it was customary to carry this sexagesimal subdivision still farther, each second consisting of 60 *thirds*, the third, of 60 *fourths*, and so on, a cumbersome method, which has happily gone entirely out of use.

17. French or Metric System. History. This system was adopted in France in 1795, during the Revolution, and was intended to furnish a uniform system of weights and measures, all of which should be based upon a single linear unit, and the various subdivisions of which should be purely decimal.

As early as 1790, Talleyrand proposed the establishment of a general system, in the construction of which all the civilized nations of the world should be invited to join, and suggested the pendulum beating seconds as an invariable standard from which to derive the linear unit. In accordance with the proposition, a party of five [1] from the Academy of Sciences was appointed to investigate the subject. This committee reported in 1791, advising that the project should be carried out, but preferred as a standard of reference the quadrant of the earth's meridian included between the equator and the pole, which was then supposed to have the same value in all longitudes, while the seconds pendulum was known to vary in length with the locality. They also suggested the employment of the pendulum as a secondary standard, from which to recover the other in case of its loss. In accordance with their recommendation, which was immediately transmitted to the Assembly, Délambre and Méchain were deputed to ascertain the length of a quadrant of

[1] Borda, Lagrange, Laplace, Monge and Condorcet.

the meridian, and for this purpose measured a meridional arc of about 9 1-2 degrees, passing through France from Dunkirk to Barcelona. This work occupied a number of years, but the Assembly made use of a value of the quadrant given by some earlier measurements, and promulgated the system in 1793, not adopting the present nomenclature, however, until 1795. At length, in 1799, an international convention was assembled at Paris, which finally decided the values of the various standards of weights and measures.

18. Linear Measure. The French arc, compared with arcs previously measured in Peru and Lapland, gave as the distance from the equator to the pole, 5,130,740 *toises*, the toise being an old French measure equal to about 2.132 English yards. It was decided that $\frac{1}{10000000}$ of this distance should be called a *metre*, and made the basis of the new system. The unit adopted in Délambre and Méchain's measurements was the *Toise of Peru*, the same that had been used in the earlier measurements of the Peruvian arc by Bouguer and La Condamine. The Act of 1799 decrees the metre to be $\frac{443296}{864000}$ of the length of the standard *Toise de Pérou*, which is an iron bar, correct at 62.25 degrees Fahr., made in 1735 under the direction of Godin. A standard metre of platinum was made from this by Lénoir. The value of the metre in English inches is 39.370432.[1]

More extensive geodetic measurements made during the present century show that the value of the quadrant obtained by the French astronomers is too small. Bessel gave as its true value 10,000,856 metres. Moreover, owing to the recently demonstrated fact that the earth is not an oblate spheroid, but more nearly an ellipsoid with three unequal axes, it follows that the value of the quadrant of a meridian varies with the longitude. Hence the basis of the French system is in reality as much a local one as is the pendulum. It should also be borne in mind that from the terms of the law of 1799 the French metre, like the present English foot, is simply the length of a legalized material scale.

The metric system is purely decimal. The larger divisions of linear measure are the *myriametre* = 10,000 metres, the *kilometre* = 1,000 metres, the *hectometre* = 100 metres, and the *dekametre* = 10 metres. Those less than 1 metre are the *decimetre* = $\frac{1}{10}$ metre, the *centimetre* = $\frac{1}{100}$ metre, and the *millimetre* = $\frac{1}{1000}$ metre.

Measures of surface and solidity are formed from those of linear dimensions, precisely as in the English system.

19. Weight. The unit of weight is the *gramme*, which was ordered by statute to be the weight of a cubic centimetre of pure water in vacuo, at 39.1 degrees Fahr. The actual standard platinum weight (by Fortin) of 1000 grammes, in the French Archives, is

[1] According to the measurements of Capt. Clarke in 1866. Two other measurements of the metre have been made; that of Kater (1821), which gave as a result 39.37079 in., and that of Hassler (1832), giving 39.3810327 in. The latter is the value which has been used by the U. S. Coast Survey. The results of Capt. Clarke are those now in use by the British Ordnance Department.

a little greater than it should be according to this law. The gramme is divided into *decigrammes*, *centigrammes* and *milligrammes*, equal to $\frac{1}{10}$, $\frac{1}{100}$, $\frac{1}{1000}$ grammes respectively. The larger weights are the *dekagramme* = 10 grammes, the *hectogramme* = 100 grammes, and the *kilogramme* = 1000 grammes. The weight of the standard kilogramme above mentioned is 15432.34874 grains, or 2.204,621,250 lbs. Avoirdupois, which gives 15.43234874 grains as the weight of the gramme.[1]

20. Capacity Measure. The unit of capacity measure is the *litre*, which is the volume of a kilogramme of pure water at 39.1 degrees Fahr., and 30 inches barometer. It was originally intended to be a cubic decimetre, but is somewhat larger, owing to the excess of weight of the standard kilogramme mentioned above. It is equal to 61.02499 cubic inches.[2]

21. Circular Measure. The French also applied the decimal system to angular measure, and a number of the circles used in the observations for determining the length of the French meridional arc were thus divided. Each quadrant contained 100 *grades or degrees*, which were again subdivided into tenths and hundredths. Hence the centesimal degree was equal to 0.9 of a sexagesimal degree. As the sexagesimal method of division was universally employed long before the metric system originated, the French division of the quadrant has never been extensively used, even by the nation which devised it.

22. Full tables of all French weights and measures employed in physical operations, together with their values in English units, will be found in the appendix to this volume.

Owing to its great simplicity, the metric system is gradually displacing all others. At the present time it is the most widely used of any, and is rapidly gaining in general favor, most of the European nations having adopted it either wholly or in part. Its employment in all commercial transactions was legalized by Great Britain, in 1864, and by the United States, in 1866[3].

Sir John F. W. Herschel has proposed the polar axis of the earth as a more logical basis than any other for a system of weights and measures, since, unlike the pendulum and the quadrant of latitude, it is not local in its nature. While this is very true, the wide acceptance of the metric system renders it undesirable to attempt a change which, even were it possible, would be chiefly of theoretical advantage. What is most desirable among different nations is some system, at once uniform and simple, both of which requisites are already supplied; the derivation of the unit is a matter of but slight practical importance. Another advantage urged by Herschel, is, however, that if the present English inch were increased by $\frac{1}{1000}$ of its total value, the polar axis of the earth would contain just 500,500,000 such inches; and that 1728 such cubic inches of water would weigh 1000 ounces Avoirdupois if the present value of the ounce were increased by $\frac{1}{18}$ of a grain. On the other hand, it is not probable that the most accurate

[1] Miller.

[2] Clarke.

[3] The student may be interested to know that the U.S. five-cent piece is 2 centimetres in diameter, and weighs 5 grammes.

value that we possess of the earth's polar axis is sufficiently near to absolute correctness to warrant a change at present.

23. Unit of Time. The unit employed in physical observations extending over a considerable time, is the *mean solar day* of 24 hours, that being the mean interval between two successive transits of the sun across the celestial meridian.

The mean interval is taken because the time elapsing from day to day between the passages of the sun across the meridian is not constant, owing to the facts that the earth's motion about the sun is not uniform, and the sun's apparent path is not in the celestial equator, but inclined to it at an angle of 23° 28′. Our ordinary clocks are adjusted to keep mean solar time. Astronomical clocks, on the contrary, are adjusted to *sidereal* or *star-time.* A sidereal day is the interval between two successive transits of the same star across the meridian,[1] and is the exact period of the revolution of the earth on its axis. It is 3 m. 56.5 s. shorter than the mean solar day. The sidereal day, like the solar, is divided into hours, minutes and seconds. When astronomical clocks are used for the estimation of time in physical experiments, it is customary to reduce their indications to mean solar time by a simple proportion. For small intervals of time the unit in common use is the *second.*

Until recently it was supposed that the period of the earth's rotation had not varied by $\frac{1}{10000000}$ of its length since 720 B. C., but an error was found by Adams in the work of Laplace, on whose calculations the above supposition was based, and he has shown that the time of revolution is diminishing at the rate of 22 s. in a century.

24. Instruments of Measurement. Physical measurements are of five kinds, according to the nature of the magnitude to be determined. Instruments for precise measurements may be classified according to their use, as follows: —

1. Instruments for the measurement of angles.
2. " " " length.
3. " " " volume.
4. " " " weight.
5. " " " time.

It will be most convenient to consider the construction of the majority of these instruments as they present themselves in the sequel, in connection with the subjects of base lines, determination of the standards of capacity measures, balances, etc., but we shall here describe a few which are of very general application. These are the *graduated circle* for angular measurement, the *vernier* and the *cathetometer*, an instrument for the precise measurement of vertical distances.

[1] More strictly, of the *vernal equinox*, but the difference is only one one-hundred and twentieth of a second in a day.

25. Instrument for measuring Angles. Angles are generally measured by circles divided at the circumference into degrees and fractions of a degree. The common *Engineer's Transit*, Fig. 3, will give a general idea of an instrument for measuring angles. It is designed to measure either vertical or horizontal angles, and therefore carries two circles, one vertical, the other horizontal, which can be seen at C and D. At AB is a telescope moving upon the same horizontal axis with D, while the whole instrument turns horizontally about an axis attached to the centre of C. Screws, S (*leveling screws*), are attached, by which the circle C can be made exactly horizontal, also screws E and F called *tangent screws*, by which a slow motion can be given to C or D. At G is seen a compass needle, which is sometimes of service, and at HK a level attached to AB, so that the telescope may be made horizontal if necessary. The telescope AB, is furnished with two cross-hairs in its eye-piece, at right angles to each other (Fig. 4). Suppose it is necessary to measure the vertical angle between two points. The observer places his eye at the telescope, focuses it on one of the points, bringing the intersection of the cross-hairs to coincide with the image of that point, and observes the reading on the vertical circle given by the index I. He then turns AB about its axis until the second point comes into the field of view, and fixes the intersection of the cross-hairs upon it in the same manner as before. Then taking the new reading given by I, the difference between it and the former reading is the desired angle. As the telescope moves horizontally about the centre of C, it evidently makes no difference whether the two points are or are not in the same vertical. Horizontal angles are measured in a similar manner by means of the horizontal circle C. There is another index opposite I, and in delicate measurements the mean of the readings of both should be taken, which will eliminate any error from eccentricity in the mounting of the divided circles.

26. Vernier. It is often necessary to measure lines and angles more closely than it is convenient to divide the scale or circles used. In such cases we may estimate fractions of one division by the eye, especially if aided by a magnifying glass, with considerable accuracy. Various other methods have been devised, one of the simplest and most generally applicable of which is the *Vernier*, so called from its inventor, Pierre Vernier, of Brussels. (1631.) To understand its construction, let AB, Fig. 5, be an enlarged representation of a scale divided into millimetres, with which we wish to read to tenths of a millimetre. Suppose that a scale DC is made by taking 9 divisions of AB (in the figure 110 to 119), and dividing this length into 10 equal parts, numbered from 0 to 10. Each division of DC will equal $\frac{9}{10}$ of a millimetre, and will be less than one division of AB by $\frac{1}{10}$ mm., so that if the 0 of DC, which is the vernier scale, be made to coincide with any division of AB, as the 750th, division 1 of DC will fall short of 751 of AB by $\frac{1}{10}$ mm., 2 will fall short of 752 by $\frac{2}{10}$ mm., 3 will fall short of 753 by $\frac{3}{10}$ mm., and so on, until the 10th division of DC, which coincides with 759 of AB. To comprehend the method of using it, let it be required to find the distance of some point S, Fig. 6, from any other point. The scale AB is laid so that its 0 shall be opposite one of the points, and the vernier is slid along AB until its 0 denoted by the arrow rests against S. The distance from the other point as indicated by the scale, is evidently 751 and a fraction millimetres, the whole number being given by AB. The fraction is obtained from CD, by counting its divisions upward from 0, until one is reached which coincides with one of the divisions of AB. The number of this vernier-division gives the fractional reading, which in this case is the number of tenths of a millimetre ($\frac{3}{10}$) between 751 and the 0 line of the vernier. For as each division of CD is less than one division of AB by $\frac{1}{10}$ mm., if the coincidence takes place as in the figure at the 3d vernier-division above 0, the space between 0 and 751 must equal 3 scale-divisions minus 3 vernier-divisions, that is $\frac{30}{10}$ mm. — $\frac{27}{10}$ mm. = $\frac{3}{10}$ mm. Hence in the instance supposed, the total distance between the two points is 751.3 mm.

From this it will be seen that in general, if we call n the number of divisions of the scale equal in length to $n + 1$ divisions of the vernier, in which case $\frac{n}{n+1}$ is the value of one division of the latter, and denote by x the distance between that scale-division which coincides with a vernier-division, and the scale-division next below the 0 of the vernier, the fractional reading is $x - \frac{nx}{n+1}$

$$= \frac{x}{n+1}. \quad (1.)$$

Verniers are applied to graduated scales and circles of all kinds, and furnish a ready means of increasing their delicacy of measurement. There are various modifications in use for particular purposes, but the principle of construction is the same in all. The smallness of the fractions which are indicated by any vernier depends upon the number of its divisions. Thus if, as in Fig. 7, 9 scale-divisions are sub-divided into 10 parts, it reads to $\frac{1}{10}$ mm. If 99 scale divisions were sub-divided into 100 parts, it would read to $\frac{1}{100}$ mm., since each vernier-division would then $= \frac{99}{100}$ mm., and one scale-division minus one vernier-division $= \frac{100}{100}$ mm. — $\frac{99}{100}$ mm. $= \frac{1}{100}$ mm. Or, generally, if x in Eq. (1) is made equal to unity, $\frac{1}{n+1}$ will be the minimum limit of reading in terms of the length of the divisions of the scale. Hence to find the limits of reading of any vernier, *divide the length of one scale division by the number of divisions of the vernier.*

27. Cathetometer. In many physical investigations it is necessary to measure the difference between the vertical heights of two or more objects to which it is difficult to apply an ordinary scale. In such cases we make use of an instrument invented by Dulong and Petit, and known as the *Cathetometer* (Fig. 7). The essential parts of this instrument are a heavy vertical pillar of metal *AB*, mounted upon a base *HK*, which is furnished with leveling screws. The pillar *AB* carries a scale *FG*, divided into millimetres. A telescope *CD* is furnished with cross-hairs, and mounted horizontally upon a frame *TN*, which can be raised or lowered by sliding it along *AB*. A screw *S* serves to give a very slow vertical motion to the telescope when desired. The frame *TN* carries a vernier *E*, and a level *L* is attached to *CD* to detect any deviation from horizontality. To measure the vertical distance between any two points, after having carefully levelled the instrument, the frame *TN* is moved until one of them lies in the axis of the telescope, which is focussed carefully, and its cross-hairs are made to coincide exactly with the point by means of the screw *S*. The reading of the scale and vernier is then taken. This done, the telescope is raised or lowered until the other point coincides with the cross-hairs, when the scale and vernier are again read. The difference between these two readings is the vertical distance between the points. That the cathetometer may give accurate results it is evident that the axis *AB* must be exactly vertical, and that the line of sight of the telescope must remain parallel to itself when raised or lowered.

28. Measure of Time. Measurements of considerable intervals of time are made with clocks or chronometers. The measurement of very minute intervals is accomplished by various forms of instruments called *chronographs.* Many of these will be described in succeeding chapters.

References.

For additional information the student may consult

Account of the Proportions of French Measures and Weights, from the Standards of the same kept at the Royal Society, by G. Graham; *Philosophical Transactions*, Vol. XLII., p. 185.

Account of Comparison of Standard Yards and several Weights with original Standard, by G. Graham; *Phil. Trans.*, Vol. XLII., p. 541.

Construction of the new Imperial Standard Pound and its Copies of Platinum, by W. R. Miller; *Phil. Trans.*, Vol. CXLVI., p. 753.

Construction of the new Standard of Length, by G. B. Airy; *Phil. Trans.*, Vol. CXLVII., p. 621. (In this article will be found an extensive list of references to articles bearing on the subject of which it treats.)

On the Correct Adjustment of Chemical Weights, by Wm. Crookes; *Chemical News*, Apr. 19, 1867.

Comparison of Standards of Leng'h of England, France, Belgium, etc., by Capt. A. R. Clarke; (London, Ordnance Survey Department) pp. 105, 163.

Tables, Meteorological and Physical, prepared for the Smithsonian Institution, by Arnold Guyot (Washington, 1858).

The Yard, the Pendulum and the Metre, a Lecture by J. F. W. Herschel, published in *Familar Lectures on Scientific Subjects* (London, A. Strahan & Co., 1867), p. 419.

The Metric System, by Charles Davies, (New York and Chicago, A. S. Barnes & Co., 1871) contains the lecture cited in the preceding reference, together with the *Report on Weights and Measures,* by John Quincy Adams, (1821) *and the Report of the University Convocation of the State of New York,* (1870).

The Metric System, with Appendix on Capacity Measures, by F. A. P. Barnard. (New York, 1872.)

Base du System Métrique, Tome III, par Jean B. J. Delambre et Pierre F. A. Méchain.

Recueil Officiel des Ordonnances et Instructions sur la Fabrication et la Verification des Poids et Mesures. (Paris, 1839.)

Atlas des Poids et Mesures dressé en Execution de l'Ordonnance Royale. (Paris, 1839.)

Rapport sur les Poids et Mesures envoyés au Gouverment des Etats-Unis, par A. Vattemare. (Measures executed by I. T. Silbermann.)

Rapport sur la Révision des Etalons en 1867 *et* 1868; *Annales du Conservatoire des Arts et Métiers,* Tome IX., No. 33, 1er Fascicule, p. 1.

Also consult references under the chapter of this work treating of the *Pendulum.*

The following *Parliamentary Sessional Papers* may be consulted for records of the legislation of Great Britain relative to the subject of Weights and Measures: —

On Weights and Measures; Parl. Papers, Reports of Committees, 1813–14, Vol. III., No. 290. Also *Rep. Committees, Jan. to July,* 1821, Vol. IV., No. 571; and *Report of Commissioners,* by Clark, Gilbert, Wollaston, Young and Kater, No. 383 of same volume.

Minutes of Evidence on Weight and Measure Bill; Rep. Committees, 1824, Vol. IV., No. 94. *Also Rep. Committees,* 1834, Vol. XIV., No. 464.

Report on Weights and Measures, and Minutes of Evidence; Rep. Committees, 1835, Vol. XIV., No. 292.

Report of Commissioners appointed to consider the Steps to be taken for the Restoration of the Standards of Weight and Measure, by Airy, Baily, Bethune, Herschel, Lefevre, Lubbock, Peacock and Sheepshanks; *Rep. Commissioners,* 1834–5, Vol. I., No. 177.

Report of Commissioners appointed to consider Steps for the Restoration of Standards; Rep. Commissioners, 1842, Vol. XI., No. 356.

Report of Committee appointed to superintend the Construction of the Parliamentary Standards, by Airy, Rosse, Wrottesley, Lefevre, Lubbock, Peacock, Sheepshanks and Miller; *Rep. Commissioners,* 1854, Vol. XIX.

Report of Committee recommending the Legalization of the Metric System, together with Minutes of Evidence; Rep. Committees, 1862, Vol. VII., No. 411.

Bill authorizing the Use of the Metric System; Parl. Bills, 1864, Vol. IV., Nos. 24, 165.

Report on the Exchequer Standards of Weight and Measure, by H. W. Chisholm, with *Notes* by G. B. Airy; *Accounts and Papers,* 1864, Vol. LVIII., No. 115.

Bill to amend acts relative to Standard Weights and Measures; Public Bills, 1866, Vol. v., No. 166.

First Report of Warden of Standards, 1866–7; *Rep. Commissioners*, 1867, Vol. xix.

Report of Commissioners appointed to inquire into the condition of the Exchequer Standards, and on the Abolition of Troy Weight, with Minutes of Evidence; Rep. Commissioners, 1870, Vol. xxvii.

On the application of the Metric System to India; Rep. Commissioners, 1870, Vol. liii., No. 225.

The legislation of the United States relative to Weights and Measures, and the reports to the Government on the construction of standards, may be found in the various *Congressional Documents*, as follows:—

Report of Committee on Fixing Standards of Weight and Measure; Reports of Committees, 15th *Congress*, 2d *Session* (1818–19), Vol. vi., Doc. 109.

Report on Weights and Measures, by J. Q. Adams, Secy. of State (Feb. 22, 1821); *Executive Papers*, 16th *Cong.*, 2d *Sess.*, Doc. 109.

Report of Committee to whom was referred the Report of the Secy. of State; Reports of Committees, 17th *Cong.*, 1st *Sess.* (1821–2), Vol. ii., Doc. 65.

Comparison of Weights and Measures of Length and Capacity, by F. R. Hassler; *Exec. Doc.* 22d *Congr.*, 1st *Sess.*, (1832.) Vol. vi., Doc. 299.

Construction of Standards of Weights and Measures; Reports of Committees, House Rep., 23d *Congr.*, 2d *Sess.*, (1835) Vol. i., Doc. 132.

Letter from F. R. Hassler, *in Rep.*, *of Secy. Treas.; Exec. Papers., House Rep.*, 24th *Congr.*, 1st *Sess.*, (1835) Vol. ii., Doc. 32.

Report on Furnishing States with Standards; Rep. Comm., House Rep., 24th *Congr.*, 1st *Sess.*, (1836) Vol., i., Doc. 259.

Ibid., Vol. ii., Doc. 449.

Report on Progress of Construction of Standards, by F. R. Hassler; *Exec. Papers*, 25th *Congr.*, 2*d Sess.*, (1837–8) Vol. xi., Doc. 14.

Report on Construction and Completion of Standards for all the States of the Union, by F. R. Hassler, (July 4th, 1838); *Ibid.*, Doc. 454.

Reports by F. R. Hassler; *Senate Doc.*, 26th *Congr.*, 1st *Sess.*, (1839–40), Vol. ii., Doc. 15; Vol. viii., Doc. 608. *Also Sen. Doc.*, 26th *Congr.*, 2d *Sess.*, (1840–1) Vol. xi., Doc. 20.

Report on Progress in Construction of Standards of Liquid and Capacity Measure, by F. R. Hassler; *Sen. Doc.*, 27th *Congr.*, 2d *Sess.*, (1841–2) Vol. iii., Doc. 225.

Reports by F. R. Hassler; *Sen. Doc.*, 27th *Congr.*, 3d *Sess.*, (1842–3) Vol. ii., Doc. 11.

Report relative to Weights, Measures and Balances, by F. R. Hassler; *Exec. Doc.*, 28th *Congr.*, 1st *Sess.*, (1843–4) Vol. iv., Doc. 94.

Reports on Weights and Measures, by A. D. Bache; *Exec. Doc.*, 28th *Congr.*, 2d *Sess.*, (1844–5) Vol. iv., Doc. 159.

Exec. Doc., 29th *Congr.*, 1st *Sess.*, (1845-6) Vol. vii., Doc. 225. *Exec. Doc.*, 30th *Congr.*, 1st *Sess.*, (1847–8) Vol. ix., Doc. 84.

Report on Construction and Distribution of Weights, Measures and Balances, and on Comparison of Foreign Standards, by A. D. Bache; *Exec. Doc.*, 34th *Congr.*, 3d *Sess.*, (1856–7) Vol. vi., Doc. 27.

Report on Weights, Measures and Balances, by A. D. Bache; *Exec. Doc.*, 35th *Congr.*, 2d. *Sess.*, (1858–9) Vol. vi., Doc. 6. (This report contains a resumé of the work relative to the construction of standards previous to

1858. Also the titles of all State Acts bearing on the subject from 1819 to 1854.)

Report of Committee appointed for Purpose of investigating the Metric System, and accompanying Bill; Acts and Resolutions of 39th *Congr.*, 1st *Sess.* (1865–6), p. 350.

For a detailed account of the various forms of *Vernier* in practical use, see *A Treatise on Land Surveying*, by W. M. Gillespie. (New York, Appleton & Co., 1867.)

For methods of graduating scales consult

Gasometry, by Robert Bunsen, trans. by H. E. Roscoe (London, Walton & Maberly, 1857); p. 25.

Traité élémentaire de Physiquè, par P. A. Daguin; Tome I., p. 22.

Cours de Physique de l'Ecole Polytechnique, par M. J. Jamin; Tome I., p. 25.

Edinburgh Encyclopedia, Article *Graduation*, by Troughton.

CHAPTER III.

Properties of Matter.

29. Matter.—General Properties. Whatever can be perceived by the ordinary operations of the senses is *matter*. It possesses two essential characteristics: (1) *extension*, or the property of occupying space, and (2) *impenetrability*, by virtue of which no two bodies can occupy the same space at the same time.

The second of these sometimes seems to be contradicted by experience. Thus when an inverted bottle is immersed in water the liquid rises to some distance inside; but in this case the air is not penetrated, it is merely compressed, so that a portion of the space which it formerly possessed is occupied by the water. A stone sunk in a vessel of water does not penetrate it, but displaces a quantity equal to its own bulk. And in all other cases in which at first sight one body seems to penetrate another, it will be found that there is really a displacement of matter.

Both the above characteristics are essential to the existence of matter. A *shadow* occupies space, but is not matter as it is not impenetrable.

30. Three States of Matter. Matter exists in three different conditions, which are known as the *solid*, *liquid* and *gaseous* states. Solids are distinguished by possessing a definite form which resists any change with considerable force. Liquids take the form of the vessel containing them, their particles not being firmly united like those of solids, but moving upon each other with the greatest ease. Gases possess this mobility of particles in common with liquids, but in addition to this their particles have a constant tendency to separate from each other, so that some ex terior effort, such as the resistance of the walls of the vessel containing them, is necessary to limit their volume. This expansive tendency may be shown by placing a bladder partly filled with air in a receiver in which we can produce a vacuum by means of an air-pump. When the external air is removed the bladder gradually swells out, owing to the expansion of the air which it contains on the removal of the external atmospheric pressure, which previously kept it under a more limited volume. As examples of the three states of matter, we may mention iron, wood and stone, among solids; water, alcohol and mercury, among liquids; and air, carbonic acid gas and common illuminating gas, among gases.

Liquids and gases are often collectively called *fluids.*

The same substance may exist in each of the three states of matter according to the external circumstances. Thus ice when heated assumes the liquid form of water, and if this is heated still more it finally assumes a gaseous state by passing into steam. Sulphur also assumes the solid, liquid and gaseous forms, successively, when heated, and many substances, as carbonic acid, laughing gas, etc., which are gaseous at ordinary temperatures, can be made to assume a liquid, and finally a solid form under the influence of great pressure and extreme cold.

31. Particular Properties of Matter.—Divisibility. In addition to the preceding there are certain particular properties which are common to all kinds of matter, though not essential to our idea of its existence.

(1.) Matter is *divisible.* We know of nothing which can not be separated into parts, these again into smaller parts, and so on. The extent to which matter is divisible will be appreciated from the following examples. Gold leaf has been hammered into leaves but $\frac{1}{25000}$ mm. in thickness. Silver wire gilded on the exterior has been drawn out to such a degree of fineness that the coating of gold was only $\frac{1}{800000}$ mm. thick. The silver core was then dissolved away by acid, leaving a gold tube of this excessive thinness. With a microscope it is possible to see a particle $\frac{1}{4000}$ mm. in diameter, so that we are able to divide gold into particles $\frac{1}{4000}$ mm. in diameter, and $\frac{1}{800000}$ mm. in thickness, each of which possesses all the qualities of the metal. Platinum wire has been drawn so fine that it took 200 metres of it to weigh 1 centigramme, its diameter being but $\frac{1}{1200}$ mm. This was done by enclosing a platinum wire in a closely-fitting silver tube, drawing out the compound wire thus made, and then removing the silver coating with acid. The thickness of a soap bubble just before bursting is but $\frac{1}{100000}$ mm. A gramme of carmine will tinge 60,000 gr. of water, or about 9,000,000 drops, so that in one of these drops there is but $\frac{1}{9000000}$ grammes of coloring material. By optical methods it is possible to detect even $\frac{1}{30000000000}$ of a gramme of soda. It is said that a single grain of musk has perfumed a room 12 feet square for several years without sensible loss of weight; and an extract of spirit of musk has given a distinct odor to 2,000,000,000 times its weight of a certain liquid. These examples might be multiplied indefinitely, but we will mention only a single additional one, drawn from the organic world. Certain microscopic plants, the *Diatoms*, are covered with a siliceous shell, weighing but $\frac{1}{30000000000}$ of a gramme. On this shell we can see striæ not over $\frac{1}{3400}$ mm. in width and thickness, thus perceiving a portion of siliceous matter, whose weight would be only about $\frac{1}{79000000000000}$ of a gramme.

32. Atoms and Molecules. The question here arises whether this division could be carried on indefinitely. Early in the history of science this was a disputed subject. Anaxagoras and Aristotle held matter to be infinitely divisible, while Leucippus taught that there were ultimate indivisible particles, to which Democritus gave the name of *atoms*. (B. C. 500.) The atomic doctrine was also taught by Epicurus. In modern times Descartes rejected the theory, while Gassendi maintained it. Up to a comparatively recent time these opinions were mere philosophical speculations resting upon no solid basis, but in later years certain laws of chemical combination, crystallography and molecular physics, have shown it to be highly probable that matter is composed of ultimate particles of inconceivable, though finite minuteness, which are physically incapable of further subdivision by any known process. These atoms are grouped together to form larger masses, called *molecules*, which may again be arranged in still more complex aggregations.

The molecules of a body are separated from each other by spaces called *pores*, which are much greater in size than the atoms themselves. Most bodies are composed of compound molecules formed by the chemical union of atoms of unlike nature to form a new substance. Hence it is convenient to distinguish between *integrant* and *constituent molecules*, constituent molecules being aggregations of similar atoms, while integrant molecules are formed by the union of the constituent molecules of dissimilar substances. Thus marble is composed of integrant molecules of carbonate of lime, while each of these integrant particles is, in its turn, composed of constituent molecules of calcium, carbon and oxygen. The exceeding minuteness of molecules renders it impossible to obtain any direct measurement of their size, but Sir Wm. Thomson has shown [1] that the mean distance between the centres of contiguous molecules of matter is probably less than $\frac{1}{1000000000}$ and greater than $\frac{1}{20000000000}$ of a centimetre.

33. Compressibility. All substances are *compressible.* On the application of external pressure to any body, its volume is

[1] Article in *Nature*, Mar. 31. 1870, p. 551.

diminished. This can only arise from the closer approximation of its molecules, which renders it evident that these are not in absolute contact. The compression of solids is noticeable in any structure supporting a great weight. The stone pillars sustaining the dome of the Pantheon at Paris, were sensibly compressed when that structure was erected upon them. In the operation of coining, the metal check upon which the impression is struck is noticeably diminished in volume. The compressibility of liquids is less evident, and until within a century they were quite generally believed to be absolutely incompressible. In 1762, Canton first showed this idea to be erroneous, and found that water was compressed about $\frac{46}{1000000}$ of its volume by a pressure of 1 atmosphere (1033.6 grammes to a square centimetre). More accurate experiments since that time have confirmed this general result. Gases are by far the most easily compressed of all substances. This may be shown by means of an air-tight piston moving in a glass tube closed at one end. The piston can be forced into the cylinder, compressing the air contained in it. On withdrawing the pressure the gas expands, forcing the piston back until it has attained its original volume. This last effect is due to the perfect elasticity of the gas, a force tending to make it return to its primitive volume. The same result occurs in the case of a liquid, and in a less degree with solids, the elasticity of the latter class of substances being in general very imperfect when compared with that of the two former.

34. Expansibility. On the application of stretching forces to bodies, or on heating them, their volume is found to increase. The expansion of solids by external forces may be shown by stretching a wire or an India-rubber tube. Expansion by heat can be rendered evident by means of a ball of iron, made of such size as to just pass through a ring of the same material when cold. If heated the ball will no longer do this, on account of its increased volume. If a quantity of liquid be contained in a tube with a bulb at one end, its level will be seen to rise in the tube when heated, and if such a bulb-tube contain air, or any other gas, a drop of liquid being placed in the tube to serve as an index, the expansion of the gas is so great that it will be apparent when the bulb is simply clasped by the hand.

35. Porosity. The fact that the molecules of any substance are not in contact, which we have already noticed, is shown to be true by the phenomena of compressibility and expansibility. The name *pores* is given to these intermolecular spaces. The porosity of substances can be proved directly in various ways. An experiment celebrated in the history of science is that first performed by Lord Bacon, and afterwards by the Academy of Florence (1661). Wishing to determine whether water was compressible, they filled a hollow silver sphere full of that liquid, and after closing it tightly flattened it under a screw-press. This operation of course diminished the capacity of the sphere. At the close of the experiment the exterior was found to be covered with a fine dew, showing that the water had been forced through the pores of the silver. The Academicians drew from this two conclusions: first, that the pores of silver were larger than the molecules of water, which is true, and secondly, that water was absolutely incompressible, which is erroneous. Other examples occur in the arts. Thus the manufacture of steel from wrought iron depends upon the absorption of a portion of carbon by the metal. It has been shown that platinum and cast iron, when heated to redness, are porous to gases. The porosity of liquids and gases is seen in the fact that in most chemical combinations the volume of the resulting compound is less than the sum of the volumes of the components. The same is true in the case of many solutions. Thus anhydrous alcohol and water mixed in the proportions of 116 to 100, sustains a diminution in volume equal to 3.7 per cent., and with other proportions there is still a diminution, though not as much as with those mentioned. The same thing occurs on mixing water and sulphuric acid. The porosity of liquids is farther shown by their ability to absorb gases. Thus water at 60° Fahr. will absorb about 720 times its bulk of ammonia gas, the resulting increase of bulk being only 50 per cent. In the case of gases, similar contractions are observed. Thus 4 volumes of nitrogen unite with 2 volumes of oxygen to form 4 volumes of protoxide of nitrogen, the resulting compound having a volume of only two-thirds that of its constituents when in an uncombined state.

In all these cases it will be seen that the particles of one kind of matter enter the pores of another, but do not penetrate the atoms, so that the statements already made relative to the impenetrability of matter are in no way invalidated.

We should carefully distinguish between the intermolecular pores, of which we have been speaking, and the larger interstices existing among the particles of solid bodies to which the name pores is also applied. These last, often called *organic* or *structural pores*, according as they occur in organic or inorganic bodies, are so large as readily to be seen with the aid of a microscope, or even with the unaided eye. Such pores may be seen in wood, and in many minerals. By slight pressure, mercury can be forced through the organic pores of a piece of wood, and *hydrophane*, a kind of agate, is opaque when dry, but on immersion in water absorbs that liquid, increasing in weight, and at the same time becoming translucent. The disintegration of rocks by frost is an effect due to the structural pores. In the wet weather of autumn, water enters them which in winter is frozen, and by its expansion causes the rock to crumble.

36. Mobility. Constant experience shows us that matter has the property of *mobility*, or capability of motion from place to place.

37. Indestructibility. Matter is *indestructible.* However greatly the external appearance of a body may be changed, not the smallest particle is actually destroyed. Thus, when a piece of wood is burnt, the only visible remainder is a small quantity of ashes, yet were we to collect the carbonic acid and aqueous vapor formed during the combustion, and weigh them, the sum of their weights, together with that of the ashes, would exactly equal the original weight of the wood. The same is true in all other cases in which matter is seemingly destroyed, owing to its assuming an invisible state.

38. Ether. The phenomena of light, heat and electricity, have led to the supposition that in addition to the three ordinary states of matter there is a fourth condition in which it may exist, and that such matter, known as the *ether*, is universally distributed, penetrating the pores of all bodies, and diffused throughout all space, existing even in the most perfect vacuum. The majority of scientists regard this ether as a substance distinct from all others, though some maintain with much plausibility that it is merely ordinary matter in a very tenuous state.[1] Ether has frequently been called an *imponderable* fluid, owing to an idea once very generally entertained that it possessed absolutely no weight. It is more probable, however, that it is ponderable, like all other matter, the impossibility of our weighing it proceeding from the fact that we cannot render any substance devoid of it, a necessary condition for ascertaining its weight.

39. Forces. The various changes which are constantly taking place in matter invariably appear to us as forms of motion, either of the body as a whole, or of its molecules. Any change of place is evidently a motion as a whole, and we shall see hereafter that all other changes, as, for example, those of temperature, are forms of molecular motion. In general, whatever produces the effect we denominate a *force*, so that we may also define a force as anything that produces, or tends to produce, motion. That the actual production of motion does not necessarily follow upon the action of a force, arises from the fact that the tendency to motion thus impressed is resisted by some other force opposed to the first. Thus, a body suspended upon a spring-balance tends to fall to the earth because of its weight; but it does not fall, since this downward force stretches the spring and develops in it a certain amount of tension, which is exactly equal and opposite to the weight of the body.

The forces acting upon matter are of two kinds, *attractive* and *repulsive forces*, the first tending to make particles or masses of matter approach, the second tending to cause them to recede from

[1] See *An Essay on the Correlation of Physical Forces*, by W. R. Grove (New York, Appleton & Co., 1865), E. L. Youmans, Editor; p. 133, *et seq.*

each other. We may also classify forces according as they act at sensible distances and between masses, or only at insensible distances and between molecules. The latter are denominated *molecular forces*. The principal forces acting at sensible distances, *i. e.*, at distances exceeding $\frac{1}{100}$ of a millimetre, are the attraction of *gravitation*, and the attractive and repulsive forces of *magnetism* and *electricity*, while those acting at insensible distances are *cohesion*, *adhesion*, *capillarity* and *chemical affinity*, together with the repulsive force due to *heat*, and certain other molecular repulsions. Gravitation acts among all bodies, causing them to tend towards one another; the forces of electricity and magnetism are manifested only under peculiar circumstances, and may be either attractive or repulsive. Cohesion acts to unite particles of the same kind of matter in a mechanical union; thus the molecules of a piece of iron are held together by this force. Adhesion causes a mechanical union of particles of different kinds of matter, as the particles of two pieces of wood are united by glue. Capillarity is a manifestation of force developed by the united action of adhesion and cohesion; and chemical affinity causes the chemical union of unlike particles of matter to form an entirely new substance. These, together with the repulsive forces which we have mentioned, will be discussed in detail hereafter.

The attractive forces acting among the molecules of a body would cause them to approach and come into absolute contact, were it not for the coincident action of a molecular repulsive force. This just balances the molecular attraction, so that the particles are kept at a certain distance apart. If they be forced nearer together the repulsive force is increased, and they tend to separate, while if the distance between them is increased by any external force, the molecular attractions cause them to tend to approach and assume their original position.

40. Polarity. Certain bodies have the property of exerting different forces at opposite ends, so that a body which is attracted by one end will be repelled by the other. The simplest illustration of this action is the case of two magnets. Suppose them to be balanced so as to move freely about a vertical axis; then, as is well known, one end of each will point towards the north. Denote by A the end of each, which assumes this position, and by B the end of each pointing toward the south. Then it will be found that if the A end of one magnet be presented to the B end of the other, an attraction will be exerted between them, while if the A end of one be approached to the A end of the second, or if the two B ends be placed near together, they will repel each other.

References.

Size of Molecules, by Sir Wm. Thomson; *Nature*, Vol. I., p. 521.
Letter on the Size of Molecules, by Sir Wm. Thomson, read before the

Manchester Lit. and Phil. Soc., Mar. 22, 1870. Published in *Nature*, Vol. II., p. 56.

Nature of Ether. See *Essay on the Correlation of the Physical Forces*, by W. R. Grove. (*Youman's Compilation.*)

CHAPTER IV.

Force and its Measure.—Laws of Motion.

41. Motion. Motion is a progressive change of position in space. We can be acquainted with none but *relative* motions, for we have no means of ascertaining the fact that a body really changes its position except by comparison with some other body not affected with the same movement. To know the *absolute* motions of any body this point of comparison must be absolutely at rest. But we can find no such point. The earth and other planets have a double motion about their axes and around the sun, the sun moves on its own axis, and also probably in an orbit around one of the fixed stars. Hence terrestrial bodies can only be at rest relatively to each other and to the earth, and as we must refer their motions to bodies which are moving themselves, the absolute motion of any point can not be determined.

Moreover, an apparent condition of rest of a body as a whole, is compatible with movements of great energy among the particles of which it is composed. We shall see as we proceed that we have reason to believe that the molecules of all substances are in a constant state of vibration to and fro, these vibrations being so minute and rapid that we can not perceive their existence except by their effects. There is no such thing, then, as absolute rest, and when the term rest is used it is to be understood as relative.

42. Velocity. The rapidity of motion is measured by the *velocity*, which is the linear space passed over in a unit of time. Thus a body moving over 10 metres in one second is said to have a velocity of 10.

43. Force.—Pressures and Impulses. A *force*, as already stated, is any cause tending to produce or modify motion. *Mechanics* is the name given to that branch of physics which treats of the laws of force in general.

Mechanical forces are commonly divided into *pressures* and *impulses*, according as the time of their duration is sensible or insensible in magnitude. Thus the weight of a body resting upon a surface is an example of a pressure, while the blow of a hammer in driving a nail is an example of an impulse. In the case of the supported weight the pressure tends to produce motion, but this tendency is resisted by the reaction of the surface on which the body rests. If this be removed motion ensues, and the body falls. Evidently the tendency of an impulse may be resisted in a similar manner.

The difference between these forms of force, however, is only one of degree and not of kind, for impulses are, in reality, only pressures acting during a very short time. The impulse given to an arrow by a bow, for example, is due to the continued action of the bow-string for a fraction of a second. The impulse given to a cannon-ball is caused by the pressure of the gases formed by the combustion of the gunpowder during the time that the ball occupies in passing through the whole length of the bore. So when motion is destroyed, as in the case of a cannon-shot fired into a wall, the body is not brought to rest immediately, but penetrates a certain distance, in proportion to the magnitude of its moving force. Since an impulse is a pressure acting during an infinitely short time, it follows that any pressure may be considered as caused by a succession of impulses repeated at infinitely small intervals.

44. Measure of Pressures. In order to estimate the magnitude of a pressure we make use of instruments called *dynamometers*. The common spring-balance (*Leroy's dynamometer*), Fig. 8, is an example. It consists of a helical steel spring fixed in a frame, and connected with an index moving over a graduated scale. On the application of a force at *B* the spring is coiled more closely, and from the amount of this coiling, as shown by the index *I*, the force is estimated. Another form of dynamometer is shown in Fig. 9. The arcs *AC*, *BD*, are connected with the arms of a steel spring *E*, which is bent more closely together in proportion as the weight suspended from *C* is greater. The magnitude of the force is read by the scale upon *BD*. Various other forms of dynamometers are used, varying in construction according to the nature and magnitude of the force to be measured. The graduation of the scales of all these instruments is performed by experiment, known weights being applied, and the corresponding positions of the index marked. If the pressure is so exerted as to cause motion, as in the case of a horse drawing a canal-boat, or a locomotive moving a train of cars, the resistance or pressure produced can be measured by interposing a dynanometer in the line of its action.

45. Measure of Moving Forces. In many cases, especially where the moving forces are impulses, it is impossible to measure them in the manner described above, and hence it is important to have some other means of estimating their magnitude. A method by which we may compare all forces is by their effect in producing or destroying motion. If, for example, we see a man throw a ball with a velocity of 5, while another man throws the same ball with a velocity of 10, the time of action of the arm upon the ball being the same in both cases, we infer that the second man exerts twice as much force as the first. This is merely reasoning from the effect to the cause, measuring the latter by the former, on the principle that the cause must be proportional to the effect.

It will here be necessary to define two terms which we shall frequently employ. The *mass* of a body is the quantity of matter that it contains. Equal masses contain equal quantities of matter. *Acceleration* is the velocity generated in a body by the action of a force during a unit of time.

The measurement of forces by the motion which they produce depends upon the following proposition, which is verified by universal experience.

(I.) *Constant forces are proportional to the products of the masses on which they act, by the accelerations impressed upon those masses.*

Hence if F, F', be two forces, and M, M', masses upon which they impress accelerations a, a', respectively,

$$F : F' :: Ma : M'a'. \quad (2)$$

Thus if two forces acting for 1 second generate velocities 1 and 2 in bodies whose masses are 3 and 5, respectively, the forces are to each other as $1 \times 3 : 2 \times 5$, or as $3 : 10$, that is, $F : F' :: 1 \times 3 : 2 \times 5 :: 3 : 10$.

If in (2) we suppose the masses of the bodies to be the same, $M = M'$, and $F : F' :: Ma : Ma'$, or

$$F : F' :: a : a' \quad (3); \qquad \text{that is,}$$

(II.) *Constant forces are to each other as the accelerations which they impress on equal masses.*

Thus if two forces acting for the same time communicate velocities 3 and 5 to a body, they are to each other as $3 : 5$, that is, $F : F' :: 3 : 5$.

If in (2) the accelerations a, a' are equal, $F : F' :: Ma : M'a$ or,

$$F : F' :: M : M' \quad (4); \qquad \text{that is,}$$

(III.) *Constant forces are to each other as the masses on which they impress equal accelerations.*

Hence if the forces F, F', acting during a unit of time, communicate equal velocities to masses 8 and 15, the forces are to each other as $8 : 15$, that is, $F : F' :: 8 : 15$.

The preceding propositions are of fundamental importance, and should be thoroughly mastered by the student before proceeding farther.

46. Momentum. The product of the massa of a body and its velocity is known as its *momentum* or *force of motion.* The momentum of a body is therefore proportional to the quantity of matter which it contains, and also to the velocity with which it moves.

If bodies whose masses are as $3 : 5$ have the same velocity, their momenta are as $3 : 5$; if the bodies are equal, and their velocities 2 and 3 respectively, the momenta are as 2 : 3; and if the masses are 3 and 5, and their velocities 2 and 3 respectively, the momenta are as $3 \times 2 : 5 \times 3$, or as 6 : 15.

The fact that momentum depends on both velocity and mass, is susceptible of numerous illustrations. A small cannon-ball moving with a high velocity is capable of doing an immense amount of damage, though its mass is comparatively small, since its rapid motion gives it a great momentum. A large ship, on the other hand, though progressing with an almost imperceptible velocity, will overcome a great resistance. Thus strong cables are sometimes snapped when the vessel scarcely seems to move, its great mass making up for its small velocity. So a slowly-moving iceberg coming in collision with a ship may destroy it.

47. Momentum a Measure of Force. Proportion (2) may be stated thus: *Forces are proportional to the momenta generated by their constant and uniform action during a unit of time, or during equal times.* The momentum impressed upon a body by any force acting for a unit of time is, then, a measure of that force.

The momentum which can be generated by the action of any force during a given time is evidently a constant quantity, whether the body acted upon be large or small. Hence if a large body is put in motion by a given force it receives a proportionally less velocity than one of smaller size. In fact, if the force F will cause accelerations a, a', in bodies of mass M, M', respectively, the momenta generated are Ma and $M'a'$. As these are equal, $Ma = M'a'$, whence

$$a : a' :: M' : M \quad (5); \qquad \text{that is}$$

the velocities[1] *impressed upon two bodies by the action of the same or equal forces during equal times are inversely as their masses.*

If a mass is already in motion, and a force acts in opposition to it, the momentum destroyed by the force is equal to that which it would generate in the same time by its action on that mass when at rest. Forces are therefore proportional to the momenta destroyed by them in equal times.

In all the preceding discussion two things have been assumed; first, that the action of the forces is *constant* and *uniform*; and, secondly, that *they act during equal times.* Any variation in the intensity of the force acting would of course render our demonstrations invalid, because we should, in reality, be comparing different forces at each moment. The forces compared must also act during equal times, because it is obvious that the longer a force acts the greater the velocity it will produce.

48. Measure of Mass. We have already defined the mass of a body as the quantity of matter contained in it. The question now arises, how shall we measure this? One method is furnished by the proportions just demonstrated. Since the momen-

[1] We can evidently write for a, a', the velocities generated by the action of the forces for a unit of time, any velocities whatever, by making our unit of time larger or smaller, as the case may be.

tum which a given force can generate in a unit of time is a constant, and the force is therefore measured by this momentum, by choosing a suitable unit of mass,[1] we may write $F = Ma$; whence $M = \frac{F}{a}$ (6). The mass of a body may then be expressed by *the constant ratio between any force and the acceleration which it will produce when acting upon that body.*

It will be shown in a succeeding chapter that any body when allowed to fall freely under the influence of gravity acquires a velocity of 9.8 metres per second, a quantity which is usually designated by the letter g. The force causing it to descend is evidently its weight W. Here a force W generates an acceleration g, and as forces are proportional to their accelerations (3), $F : W :: a : g$ (7), whence $\frac{F}{a} = \frac{W}{g}$, or $M = \frac{W}{g}$ (8). The mass of any body is therefore its weight divided by the velocity which it would acquire by falling freely for one second. This quotient is a constant, for if W varies from any cause, g, the acceleration produced by it, must also vary in the same ratio (3), so that the value of $\frac{W}{g}$ will remain unchanged.

This mode of indicating the mass of a body is the one which has usually been followed in modern treatises upon the subject, but some inconveniences, the nature of which we shall presently explain, have more recently led to the use of another method.[2]

The weight of a body may be considered under two aspects: First, as denoting the *force* with which it is drawn towards the earth; and secondly, as denoting the mass of the body as compared with the mass of an arbitrary standard, such as the gramme or pound.

Now if by the weight of a body stated in grammes or pounds, we understand the force with which it is attracted towards the earth, that is if we consider the standard unit of weight to be a unit of force, we cannot express the mass of a body in grammes or pounds, because, as we shall see hereafter, the weight of a given quantity of matter, and hence the quantity of matter contained in a given weight is not the same at all places upon the earth's surface, so that a body of constant mass will possess different weights at the equator and poles. But if by the weight thus expressed we understand merely the quantity of matter contained in the body, compared with the quantity of matter contained in the arbitrary standard of weight of the French or English systems;

[1] For the momentum Ma being *proportional* to the force F, the latter must equal a *constant* multiple of the momentum, whatever be the unit of mass chosen, and if this unit be taken so that when $F = 1$ and $a = 1$, M also equals 1, or $F = Ma$.

[2] See *A Treatise on Natural Philosophy*, by Thomson and Tait, Vol. I., p. 166.

that is, if by a gramme or a pound we understand a quantity o matter equal in mass to the standard gramme or pound, we can express the mass of a body in such units, which, though arbitrary, are yet definite and invariable. In this sense it is perfectly correct to speak of the mass of a body as being so many grammes.

Now in common life this use of weight as an estimate of mass is far more general than its use as an estimate of force. In strict language, to be sure, weight is the downward tendency of a body, but we ordinarily employ weights "for the purpose of measuring out a definite quantity of matter; not an amount of matter which shall be attracted to the earth with a given force." Hence our standards of weight are primarily intended as units for measuring mass, and it is a secondary application by which we use them to estimate forces. The latter method of measuring mass is therefore far more simple than the former.

49. Unit of Mass. According to the first method of measurement the unit of mass is the mass of g times the standard unit of weight. For if the unit of mass is so chosen that $M = \frac{W}{g}$, it is evident that if $M = 1$, W must be numerically equal to g. Hence the unit of mass is the mass of g grammes, or g pounds of matter.

The objection to this mode of estimating mass is now apparent. Since the quantity g is a variable, the unit of mass is also a variable quantity, and it is exceedingly desirable to have some unit which shall be invariable. This is furnished in the second method of estimating mass, as in that case the standard gramme or pound is taken as the unit. This gives us an absolute unit of mass, which can be obtained in no other manner. This unit was first brought into general use by Gauss.

50. Unit of Force. The unit of force is the force which, by acting upon a unit of mass for a unit of time, generates a unit of velocity.

The unit of force, according to the first system, is the gramme in French, the pound in English measures. For the unit of mass in that system is g times the unit of weight, and the force with which this is attracted towards the earth is g grammes or g pounds, which in 1 second generates a velocity of g feet, or metres. Hence, as forces are proportional to the accelerations which they produce in equal masses, the force which would generate an acceleration of 1 unit is equal to $\frac{1}{g}$ of that which generates an acceleration of g units, that is, to 1 gramme, or 1 pound. This is called the *gravitation unit of force*, and the system of measurement of masses and forces derived from it is called the *gravitation system* of measurement.

According to the second of these methods of measurement, the unit of mass being a standard gramme or pound, the unit of force is the force which, acting upon a national standard unit of mass for a unit of time, generates a unit of velocity. This unit of force is numerically equal to the unit of mass divided by the acceleration of gravity, that is, to $\frac{1}{g}$ grammes, or $\frac{1}{g}$ pounds, the value of g being the acceleration at Paris or at London, according to the system used. It is known as Gauss' *absolute unit of force*, and the system of measurement of masses and forces based upon it as the *absolute system* of measurement.

To obtain a clearer idea of the value of the absolute unit of force, we must know the numerical value of g. The value of g at London is 32.1889 ft.; and at Paris 9.8087 metres. Hence the British absolute unit of force is equal to the weight of $\frac{1}{32.1889}$ lbs. at London, and the French absolute unit is equal to the weight of $\frac{1}{9.8087}$ grammes at Paris. That is, in round numbers, the British absolute unit of force is equal to the weight of about half an ounce, and the French unit to about the weight of $\frac{1}{9}$ of a gramme.

We may evidently define the French unit of force as *that force which, by acting upon a mass of* 1 *gramme for* 1 *second, generates a velocity of* 1 *metre*, and the British unit as *that force which, by acting upon a mass of* 1 *pound for* 1 *second, generates a velocity of* 1 *foot.*

To transform forces expressed in gravitation units to absolute units, we must evidently multiply their numerical value by g. Thus a force of 10 grammes is equal to 10×9.8087 French absolute units of force, and a force of 10 lbs. is equal to 10×32.1889 British absolute units.

The student will notice that the equation $W = Mg$, which according to the first system of mass measurement denotes the weight of a body of mass M, in the second system denotes the number of absolute units of force in the downward tendency W, caused by gravity.

51. Representation of Forces by Right Lines. In many problems it is convenient to use a graphical method of representing forces. A force is defined when its magnitude, direction and point of application are given. Hence we may represent the relative magnitude of forces by straight lines, whose lengths bear the same relation to each other as the numerical values of the forces themselves, while the directions of these lines may indicate the direction of the forces, and the point from which the lines are drawn the point of application. Thus two forces of 1 and 3 kgr., applied at a single point and inclined 20° to each other, would be represented by lines AB, AC, Fig. 10, of lengths 1 and 3 units respectively, drawn from a point A, and making an angle $BAC = 20°$.

52. Three Laws of Motion. Law I. The elementary principles relating to the phenomena of motion and force were arranged by Newton under the laws often known as the *Newtonian Laws of Motion.*

LAW I. *A body at rest continues in that state, and a body in motion proceeds uniformly in a straight line, unless acted upon by some external force.*

This law is directly deducible from our fundamental ideas regarding cause and effect. Since every effect requires some cause to produce it, it follows that a body unaffected by any force must remain in the state in which it already exists. A body at rest has no power to put itself in motion; a body in motion has no power to bring itself to rest, or to deviate from its path, but must continue to move in a fixed direction.

The law is also verified by all experience. The incapacity of bodies at rest to change their condition is so obvious as to need no illustration. But the case when they are in motion is not always so clear. Thus a body thrown horizontally does not move uniformly in the line of projection, but proceeds with a gradually diminishing horizontal velocity, until it comes to rest on the ground. This decay of motion seems to contradict the second portion of the law. But a closer examination shows that there is here a force acting upon the body constantly tending to destroy its motion, viz., the resistance of the air; while another force, the attraction of gravitation, continually draws it towards the earth. Again, a ball rolled upon a road does not move on indefinitely, but soon slackens its velocity, and is finally brought to rest. This is no more an exception to the law, however, than the other case, for it is the friction of the ball upon the road that causes its loss of motion. If the friction is diminished the distance to which the ball will roll is proportionally increased: on a common road it stops very soon, on a smooth bowling-green it rolls much farther, while on a sheet of clear ice it goes a very long distance before stopping. In cases where the friction and resistance of the air are reduced to a minimum, we obtain a very long continuance of motion. Thus a nicely-balanced wheel, moving on fine, well-oiled bearings, on being set in rotation will revolve for a very long while. Again, an ordinary clock-pendulum, if detached from the rest of the works and swung, will come to rest after a short interval, but some pendulums of very delicate construction, moving with exceedingly little friction will swing in the air for nine or ten hours, and in a vacuum for twenty-four hours. Since, then, whenever a diminution of motion occurs, we are able to find forces causing it, and since the condition of uniformity of motion is approached in proportion as we do away with these forces, we are justified in concluding that could we eliminate them altogether the loss of motion would entirely disappear. So in all cases of motion in curved lines, as in those just cited of the wheel and pendulum, we find forces at work which deviate the body from the rectilinear path it would otherwise pursue. We therefore conclude that the law, as stated, is true in all cases.

Still further, the consequences deduced from other laws of matter, supposing this law to be true, agree with the results of observation, which could not be the case if the law were false. Thus in Astronomy, the methods used in predicting eclipses assume that the motion of the earth and moon is not altered except by the action of some external force, and as the observed and computed times agree, we are justified in asserting the truth of our proposition. The same remark applies to the remaining laws of motion.

53. Resisting Medium in Space. The question will be asked here, "Is there any example of permanent motion in nature?" The revolution of the planets around the sun offers the nearest approach to perpetual motion with which we are acquainted, but even in this case there is evidence of the existence of a resisting medium acting to slacken their velocities. This medium is so light that its effects are only perceptible in the retardation of a single comet (Encke's), but if it exists it must nevertheless act in the same manner upon all bodies circulating about the sun. The reality of the existence of a resisting medium, though generally believed, is discredited by as high an authority as Sir J. F. W. Herschel, who explains the retardation of Encke's comet in a different manner.[1]

54. Phenomena illustrating the First Law of Motion. The truth expressed in Law I. is the principle of the *Inertia of Matter*. The term *inertia* in strict language means simply the inability of matter to change its state except under the action of some force. It is also universally used in a somewhat different, though analogous sense, as denoting *that property of matter because of which a definite force is necessary to produce a given change in the existing state of a mass*, which is simply another mode of expressing the general idea of momentum.

By means of the first law of motion in connection with the principles of momentum, numerous familiar phenomena are readily explained. The apparent resistance experienced on setting a body in motion, or on stopping a moving body, is a consequence of its inertia. A person riding on horseback at a rapid rate is thrown forward if the horse suddenly stops, because the momentum of his own body carries him onward in the direction in which he was moving. The same thing occurs to the passengers in a railway train stopped by a sudden application of the brakes. If, on the contrary, the car is started quickly they are thrown backward, the motion of the car not being immediately communicated to them. It is because of the inertia of matter that the effects of the collision of trains of cars are so terrible. The locomotive is suddenly brought to rest, while the cars continue to move, thereby piling one upon another, and causing a general destruction. For a like reason when a vessel moving at full speed strikes a sunken reef, the spars upon her deck and even the sailors are sometimes shot violently forward over the bow.

[1] See Herschel's *Outlines of Astronomy*, 11th Ed., 1871, §§ 577, 570.

"Coursing owes all its interest to the intuitive consciousness of the nature of inertia which seems to govern the measures of the hare. The greyhound is a comparatively heavy body moving at the same or greater speed in pursuit. The hare *doubles*, that is, suddenly changes the direction of her course, and turns back at an oblique angle with the direction in which she had been running. The greyhound, unable to resist the tendency of its body to persevere in the rapid motion it had acquired, is urged forward many yards before it is able to check its speed and return to the pursuit. Meanwhile the hare is gaining ground in the other direction, so that the animals are at a considerable distance asunder when the pursuit is recommenced. In this way a hare, though much less fleet than a greyhound, will often escape it."[1]

We see a practical application of the first law of motion in the method often used of fixing an axe or hammer firmly on its handle. The tool being placed in a vertical position, with the head uppermost, is moved rapidly downward, so that the end of the handle strikes against some solid object. The motion of the handle is stopped, while the head moves on, thus fixing itself firmly.

The principle under consideration may also be illustrated by a very simple experiment. Let a smooth card be balanced on the tip of one of the fingers. On this place a somewhat heavy coin. If, now, a quick, horizontal blow be given to the edge of the card with the fore-finger of the other hand, it flies from under the coin, leaving this poised upon the finger. The slight force of friction exerted by the moving card upon the coin, is not sufficient during the short time of its action to impress the motion upon the latter. If the card be pushed slowly the coin moves with it.

55. Time required to produce Change of State in a Body. From these examples, especially the last, we see that if a force acts upon only a few points of a body, a sensible time is required to transmit its effects to every portion of the mass. Several curious phenomena are explained by this fact. A light stick supported by placing its ends upon the edges of two wine glasses will be broken by a quick blow, without affecting its supports, while a less sudden stroke will be liable to break them. In the former case the stick is snapped before sufficient time has elapsed for the blow to be communicated to the glasses. A rifle ball fired through a pane of glass cuts out a round hole, because it progresses too rapidly to allow the motion to be impressed upon any of the particles of the glass, except those immediately in its path, before it has passed entirely through. A slowly-moving bullet, or a stone thrown against the pane, cracks it in every direction. For the same reason the greatest damage in naval combats is caused by shot moving with a comparatively slow motion, as they cause far more splintering of the timbers than a swifter projectile. In like manner a cannon ball may be fired through a partly open door, scarcely moving it on its hinges. A candle fired from a musket will penetrate a board without being greatly crushed. A certain time is required to change the form of a body by crushing, and as this is greater than the time required to overcome the cohesion of the fibres of wood composing the board, the candle goes through before mnch change in its form can take place.

It is because of the time necessary to generate motion in a body that when a horse attempts to start a heavy load by a sudden pull, some part of the harness is liable to give way, while the load remains unmoved, though

[1] Lardner, *Treatise on Mechanics*, p. 39.

a slow and steady pull would have put it in motion. In the case of long trains of heavily loaded cars, a great gain ensues from the slight movement allowed by the couplings which unite them to one another. That the train may be set in motion it is necessary to overcome the friction of all the wheels resting upon the rails. The locomotive easily exerts force enough to put the first car in motion, as this can be fairly started before the second is moved at all, because of the slackness of the connecting coupling, and so on for each car in succession. The train is thus put in motion in parts, the inertia of the cars already moving aiding the engine in starting the remainder. Were the whole train rigidly connected, it would be impossible for an ordinary locomotive to cause it to move. When a person falls from a considerable height upon a rock he is severely hurt, but if upon a soft bed he is uninjured, because in the latter case the stopping of the motion is performed gradually, so that the violent shock otherwise produced is obviated.

56. Matter being purely passive, it follows that if any force, however small, act upon a body, it must produce a proportionate amount of motion, supposing the body to be absolutely free to move. In practical cases there is a certain resistance to motion offered by friction and other causes, which must be overcome before movement can take place. The least force in excess of this will start the body. Hence it is a general law that in the absence of resisting actions *the smallest force is capable of moving the largest body.* As an illustration of this it is stated[1] that in calm weather and smooth water a very large ship can gradually be put in motion by the efforts of a child pulling at a rope attached to the bow. The effect of the force in producing momentum is of course exactly the same as if it were expended in giving a rapid motion to a small body.

57. Resistance of Media. The resistance offered by a fluid to the motion of a body is due to the pushing aside of its particles by the moving mass, which thus loses a portion of its momentum. The denser the fluid, that is, the heavier its particles, the more resistance it offers to movement in it. Thus a delicate pendulum swung in the air will oscillate for some hours before being brought to rest, while if swung in water it will move for only a few minutes.

58. Law II. *If several forces act upon a body simultaneously, each one of these produces the same effect in magnitude and direction as if it acted alone. This principle is known as the Law of the Independence of Motions.*

The truth of the second law of motion may not appear at first sight. The effect of mechanical forces is to produce motion along their lines of action, and the question will be asked, "How can several motions in different directions coëxist in a body?" The difficulty disappears, however, if it is recollected that all movement in a body is estimated by reference to some point not possessing the same motion. The *absolute* motion of a mass can evidently have but a single direction, but *relatively* to certain points a body may have several simultaneous motions. For example, a man travelling from southeast to northwest is at the same time moving toward the north and towards the west, so that after the lapse of an interval

[1] Leslie, *Elements of Natural Philosophy;* Vol. I., p. 30.

of time he will have passed over a certain number of miles in a northerly, and a certain other number of miles in a westerly direction.

59. Illustrations of Second Law of Motion. Various illustrations may be cited in proof of this law. The muscular exertion put forth in walking a given distance upon the deck of a steamer is the same, whether the boat is at rest or in motion. If the vessel moves at the rate of five miles an hour with reference to some point on the shore, and a person walks over the deck in the same direction and at the same rate, his velocity relative to that point will be ten miles an hour, and relative to the vessel five miles an hour. If, on the contrary, he walks in an opposite direction, his velocity relative to the vessel will be five miles, but relatively to the shore he will be at rest. Again, imagine a boy sliding upon a moving cake of ice in a direction at right-angles to its line of progression. The boy will be affected with both motions, moving down the stream as rapidly as if he were not sliding, and sliding as rapidly as if the cake of ice were at rest. So two persons standing on the deck of a steadily-moving vessel, can toss a ball to each other as readily as if the boat were at rest. To take the illustration first used by Galileo, a person on shipboard can write with ease, while in his pen coexist the motions caused by the hand, the swaying of the ship and its progressive motion, and the axial and orbital revolutions of the earth. But as the paper possesses all these movements except the first, the letters are traced exactly as if only this motion existed. A ball dropped from the mast-head of a moving ship will strike the deck at the foot of the mast, because the forward motion possessed by it in common with the vessel, is in no way diminished by the downward motion communicated by gravity. A heavy body falling from a balloon which is moving rapidly in a horizontal direction, "during its descent partakes of the balloon's motion, and until it reaches the earth is always seen perpendicularly under the car." A body dropped from the summit of a lofty tower, or allowed to fall down a deep mine-shaft, strikes the ground a little *east* of the vertical passing through the point from which it starts. This is because the earth's revolution on its axis gives to all objects a motion from west to east, the velocity being greater in proportion to their distance from the axis. Hence the easterly velocity at the summit of a tower is greater than at its base, and a body starting from that point, while falling downwards, must at the same time move toward the east more rapidly than the base of the tower, and hence strikes the ground a slight distance to the east of the vertical. The same explanation applies to the fall of a body down the shaft of a deep mine. A marked deviation of this kind has been shown by experiments made in some of the mines of Saxony. In the case of circus-riders who leap through hoops from the back of a horse running at full speed, and again alight on the animal, it is simply necessary to leap upward and not forward, because their motion in common with the horse carries them onward through the air with the same rapidity as if no leap had been made. Finally, a cannon-ball exerts the same mechanical effect, whether it is fired towards the east, west, north or south, though its velocity with regard to any fixed point in space is vastly different in these different cases, owing to the rotation of the earth. Thus suppose the ball to be propelled from the cannon with a velocity of 470 m. per second. At the equator the rotary motion of the surface of the earth is also 470 m.

per second; hence, if at any station situated in that great circle the ball be fired towards the east, its velocity relatively to the centre of the earth is 470 — 470 m., or 0 m. That is, in the latter case the ball comes to rest relatively to the earth's centre, while the surface moves on, leaving it behind. To us travelling east with the surface, the ball appears to have a westerly velocity of 470 m. per second. The destructive effect of the ball would manifestly be the same in either case.

60. Law III. *Action and reaction are equal and in opposite directions, or the actions of bodies on one another are equal and opposite.*

This law is established solely by induction. To exemplify its truth it will be sufficient to examine a number of phenomena of different classes, and note its applicability to all.

61. Illustrations. A magnet exerts a force of attraction upon a bit of iron, but the iron at the same time attracts the magnet with an equal force. If the magnet is fixed, and the iron free to move, the latter will approach the former. If the iron is fixed and the magnet free, the magnet will move towards the iron. If both are free they will approach each other. And this action is equal as well as mutual and opposite. For it is found by experiment that in the last case the velocity imparted to the iron is as much greater than that imparted to the magnet as its mass is less. That is, if M, M' are the masses; and v, v' the velocities, we shall have $M : M' :: v' : v$. Hence the momenta are equal, for $Mv = M'v'$, and therefore the forces generating these momenta are also equal. The *action* of the magnet, then, is equal to the *reaction* of the iron.

It is to be observed that the terms action and reaction do not denote two forces, but are simply convenient terms by which we express the mutual and opposite actions of the same force. The magnet and iron are drawn together by the single force of their mutual attraction. So when the elasticity of an uncoiling spring pushes apart two bodies connected with it, there is really but one force at work, which moves them in opposite directions, though it may sometimes be convenient to consider the motion as caused by two equal and opposite forces acting from the middle towards the end of the spring. If the masses of the bodies are in any given ratio, as 1 : 2, for example, the velocity impressed on the smaller will be twice that of the latter, so that the momenta will be equal; thus again verifying the law under consideration.

In the case of a bullet fired from a gun, the force of the exploding powder propels the ball forward at a higher velocity. At the same time, however, it forces the gun itself backward, causing the recoil or *kicking* of the piece. If the force of the recoil be measured, it will be found that the momentum of the gun equals that of the bullet. When a man-of-war fires a broadside, the whole vessel sways in the opposite direction. Many other examples may be mentioned. In the exercise of rowing, the water is pushed backward, while the boat moves forward. A person leaping from a small boat pushes it away from him, as he springs away from it. It is because of this principle that a person cannot lift himself into the air by pulling at his boot-straps. The upward pull on the straps (action) is just balanced by the downward push (reaction) of the feet. A case is related of a gentleman who undertook to propel a boat by erecting a large bellows at the stern, and blowing against the sails,[1] the attempt resulting in a signal failure, because the reaction of the current of air as it issued from the bellows neutralized its action upon the sail.

It follows from the law of the equality of action and reaction, that if a body in motion strikes upon one at rest, the shock is the same for both, because the loss of momemtum which the moving body sustains affects it in the same manner as if being at rest it was acted upon by an equivalent force in an opposite direction. If two bodies moving in opposite directions impinge upon one another, the same force is exerted as if one while at rest was struck by the other moving with a momentum equal to their united momenta. This explains why the shock is so violent when two vessels moving in opposite directions run foul of one another. In the case of bodies thus impinging upon each other, though action and reaction are equal. the weaker will evidently be the more injured by the impact. The fist of a pugilist sustains as great a shock as that part of the opponent's body which it strikes, but is not injured, because from its structure it is fitted to endure the shock. But if by accident fist meets fist, one person feels the blow as much as the other.

62. History of the Laws of Motion. Simple and fundamental as are the three laws of motion, it was not until the beginning of the 17th century that they were understood. Kepler, with all his success in astronomy, was ignorant of the principles of inertia. He supposed that moving bodies, if left to themselves, would come to rest, and therefore imagined a constant force acting upon the planets to keep up their velocity. Galileo, during his earlier researches, thought that the only naturally uniform motion was that performed in a circle, but in his *Dialogues on Mechanics*, published in 1638, he gives a correct statement of the first law, though it is not made sufficiently comprehensive in its application. The law, in a general form, was announced by Galileo's pupil, Borelli, in 1667.. The truth of that portion of the law relating to the tendency of bodies to move in right lines was recognized by Bendetti, as early as 1585. The principle of the independence of motions was also announced in the *Dialogues on Mechanics*, but its complete demonstration resulted from the establishment of the laws of the motion of the earth by the astronomers of the 17th century, foremost among whom was Sir Isaac Newton. The principles of momentum underlying the third law were known to Galileo, but the laws of impact of bodies, as related to changes in momentum (a knowledge of which was evidently necessary to a general statement of Law III.), were first correctly stated by Wren, Wallis and Huyghens, in papers communicated to the Royal Society about 1609. The terms in which it is now so frequently expressed, "Action and Reaction are equal and opposite," are those used by Newton.

References.

For information upon the subject of Units of Force, see

Treatise on Natural Philosophy, by Sir Wm. Thomson and Peter G. Tait; (Oxford, 1867) p. 166.

Numerous examples of the application of the laws of motion will be found in

Elements of Physics, by Neil Arnott.

[1] Arnott's *Elements of Physics*.

A Treatise on Mechanics, by Henry Kater and Dionysius Lardner.
Handbooks of Natural Philosophy and Astronomy, by Dionysius Lardner; *First Course.*

For further information upon the subject of a resisting medium see

Outlines of Astronomy, by Sir J. F. W. Herschel; 11th English Ed., §§ 577, 570.
Reports on Observations of Encke's Comet during its Return in 1871, by Asaph Hall and Wm. Harkness (Washington, 1871); p. 33.

CHAPTER V.

Composition and Resolution of Motions and Forces.

63. Statics and Dynamics. The science of mechanics is divided into *Statics* and *Dynamics*. Statics treats of balanced forces, or forces in equilibrium; dynamics of the action of forces in producing motion.

The demonstration of the elementary principles of statics requires the preliminary consideration of the composition of motions; that is, of the laws determining the path described by a body in which several motions coexist. The fundamental theorem upon which they all rest is the law of the independence of motions.

64. Parallelogram of Motions. *If a particle be simultaneously impressed with two uniform motions which separately would cause it to describe the adjacent sides of a parallelogram in a given time, it will describe the diagonal of that parallelogram in the same time, and uniformly.*

Suppose a particle at A, Fig. 11, to possess at the same time two uniform motions, one of which would carry it over the line AB, the other AD in a given time. It will move uniformly over the diagonal AC in the same time.

For by Law II. (§ 58, p. 41) each motion takes place as if the other did not exist, hence the motion along AD cannot affect the movement of the particle in a direction parallel to AB, and it will therefore proceed as far in that direction as if it were not impelled along AD; that is, at the end of the given time it must be found somewhere on BC. Also the motion along AD produces its full effect, as if the particle were subject only to it, and hence causes the particle to proceed as far in a direction parallel to AD as if it were not impelled along AB. The particle must therefore be on the line DC at the expiration of the given time; and since it is also on BC it will be found at their intersection C.

This motion is entirely performed in the diagonal AC. For let Ab, Ad, be the distances which would be passed over in any fraction of the whole time if the motions took place separately. Then since, by supposition, the motions over AB, AD are uniform,

Time of passing over Ab : *Time over AB* : : *Time over Ad* : *Time over AD*; or, as the times are proportional to the spaces described,

$$Ab : AB :: Ad : AD.$$

Hence the parallelograms $Abcd$, $ABCD$ are similar, and the point c lies in the diagonal AC. But, by the first part of the demonstration, c will be the position of the body at the end of the given fraction of time; and as this is true whatever be the time chosen, the whole path must lie in AC.

The motion in AC is uniform. For, by similarity of triangles, $Ac : AC :: Ad : AD$.

But $Ad : AD$: : *Time over Ad* : *Time over AD*,

And *Time over Ad* : *Time over AD* : : *Time over Ac* : *Time over AC*.

Whence *Time over Ac* : *Time over AC* : : $Ac : AC$; that is, *the spaces described are proportional to the times occupied in describing them*, in which case the motion must be uniform.

Also *the velocity in the diagonal is to the velocity in either side, as the length of the diagonal is to the length of that side*, since each is traversed in the same time.

65. Triangle of Motions. It follows from the preceding proposition that *if a particle possess simultaneously two uniform motions which, if taking place in succession, and each continuing for the same interval of time, would cause it to describe two sides of a triangle taken in order, it will describe the third side in the same time.* For the motions which carry a particle over AD, AB, Fig. 12, if applied at A, would by acting successively, carry it over AD, DC. Hence the effect of their simultaneous occurrence is the same in either case, producing a motion in AC, the diagonal of $ADCB$, and third side of the triangle ADC.

66. Component and Resultant Motions. The motions which are thus combined are generally known as *component* or *elementary motions*, while the single motion due to their combination is called the *resultant motion.*

67. Illustrations of Composition of Motion. The composition of motions may be illustrated by placing a ball upon a level square table, and communicating to it two equal impulses along the sides; it will be found to move in the diagonal. If a white ball suspended before a blackboard have a horizontal and a vertical motion communicated to it simultaneously by means of cords, it will be seen to move in an oblique direction. The apparatus shown in Fig. 12 illustrates these principles very clearly. Within a rectangular frame, $ABCD$, slides a second frame, $EFGH$. A white disc K slides with an easy motion upon a rod FN, attached to the inner frame. A cord attached to the disc runs over a pulley P at the upper extremity of the rod, and is fastened to the outer frame at B. Now it is clear that if the frame $EFGH$ be drawn in the direction of the arrow, the disc K is carried horizontally with the rod FN, while at the same time it moves vertically over FN, owing to the action of the string which is fastened at B. Under the combined effect of these two movements, the disc will be seen to traverse the diagonal GL.

Practically, we notice the composition of motion in the case of a boat rowed across the river, while it is at the same time carried down the stream by the current. The boat moves obliquely, reaching the opposite shore in the same time as if there were no current, but at a point lower down the stream. If the boatman wishes to reach a point directly opposite he must evidently head up the stream, so that the resultant of the combined motions of the boat may lie in a straight line joining the two points. For example, suppose it is wished to cross a stream directly from A to C, Fig. 13, the current being sufficiently strong to carry the boat a distance AD, while the boatman can row it over AB. The boat must be rowed in the line AB, in which case the motions in AB and AD give a resultant motion in AC. The time of crossing is clearly that which it would take to pass from A to B in still water.

68. Polygon of Motions. *If a particle possess simultaneously any number of uniform motions, which occurring successively, and each continuing for the same interval of time, would cause it to describe all the sides but one of a polygon, it will describe the remaining side in the same time, and with a uniform motion.*

Let a body at A, Fig. 14, be impressed with uniform motions, which, if taking place in succession, would cause it to move over the sides AB, BC, CD of the polygon $ABCD$, the time of describing each side being the same. Then by the theorem of the *triangle of motions*, the combination of the movements parallel to AB, BC, gives a resultant motion over AC. Now combine this resultant with the remaining elementary motion parallel to CD. The resultant of these, which is also the resultant of all three motions, is over the line AD. But AD is the fourth side of a quadrilateral, of which AB, BC, CD, are the other three sides. If a fourth elementary motion were present it could be combined with AD, the resultant of the first three (in which case the resultant would be the fifth side of a pentagon), and so on, indefinitely. Hence the proposition is general.

The uniformity of the resultant motion is proved precisely as in the case of two combined movements.

69. Kinematics. The science which treats of the relative motions of bodies considered independently of the causes producing them, is called *Kinematics.*

70. Composition of Forces. By the aid of the preceding propositions we shall be able to demonstrate the laws of the effect of several forces when acting simultaneously.

Just as elementary or component motions combine to produce a single resultant motion, *component forces* combine to produce a single *resultant force.* The fundamental theorem relative to the composition of forces is the following.

71. Composition of two Forces. — Parallelogram of Forces. *If two forces are represented in magnitude and direction by the adjacent sides of a parallelogram, their resultant will be represented in magnitude and direction by the diagonal.*

Let F, F', be two forces represented in magnitude and direction by the sides AB, AD, of the parallelogram $ABCD$, Fig. 15. Then will their resultant R be represented in magnitude and direction by the diagonal AC. For suppose F, F' to act separately upon a particle at A to produce motion in it. Since forces are proportional to the accelerations which they produce in the same mass, or equal masses (p. 33, Eq. 3), the particle will traverse AB under the sole action of F, in the same time that it will move over AD under the action of F'. The velocities generated by the forces would then be such as to cause the particle to describe two adjacent sides, AB, AD, of a parallelogram in equal times. Hence if both forces act simultaneously, the combination of these velocities will produce a resultant motion represented by AC, which measures the force producing it. But this is evidently R, the resultant force due to the combined action of F and F'. Hence F, F', and R must bear to each other the same relations in magnitude as AB, AD and AC.

AC also represents the *direction* of the resultant because motion can take place only in the line of the force producing it.

72. Triangle of Forces. Since $AB = DC$, and AD, DC, AC, form three sides of a triangle, it is evident that the same forces which are represented in magnitude and direction by AD, AB, AC, two adjacent sides of a parallelogram and its diagonal, are also represented by AD, DC, AC, three sides of a triangle ADC. *Hence, if two forces are represented in magnitude and direction by two sides of a triangle, the third side will represent their resultant.*

73. Particular Cases of Combination of Forces. It is evident that if the angle $A = 180°$, the forces F, F', act in direct oppositition, and the resultant will equal the difference of the components, that is, $R = F - F'$ (9). On the other hand, if $A = 0$, they act together, and the resultant is the sum of the components, that is, $R = F + F'$ (10).[1] Hence if we combine these propositions, and distinguish all forces acting from left to right by the sign $+$, and those acting from right to left by the sign $-$, *the resultant of any number of forces acting in the same straight line equals the algebraic sum of the components.* If R be this resultant, and ΣF the sum of the components,[2] $R = \Sigma F$ (11).

[1] Geometrically, this follows from the *Parallelogram of Forces*, because if the angle DAB (Fig. 15) $= 180°$, AC becomes equal to $AB - AD$; while if $DAB = 0°$, AC becomes equal to $AB + AD$.

[2] The Greek letter Σ is frequently used to designate the sum of any finite number of quantities expressed by a symbol placed after it.

74. Illustrations of the Composition of two Forces. The composition of two forces may be illustrated by the case of an arrow fired from a bow. The bowstring ACB, Fig. 16, is brought into a state of tension by the bent bow, hence each half of the string exerts a pull at C towards the extremity of the bow to which it is attached, so that the forces acting upon the arrow lie in the lines CE, CG. Representing their magnitude by CE, CG, the diagonal CF represents the magnitude and direction of the resultant. Hence the arrow, when the string is released, moves forward in the line CF under a force bearing the same relation to the tension of the string in either direction that CF bears to CE or CG.

75. Solution of Problems. Problems relating to the composition of forces are readily solved by the principle of the *Parallelogram or Triangle of Forces*, either by graphical construction, or by the application of trigonometry. Let F and F' be the given components, α the angle made by their lines of action, and R the required resultant. (1.) *Graphical Solution.* — To solve the problem graphically, lay off $AB = F$ units of length, and $AD = F'$ units (Fig. 17), making $BAD = \alpha$. Complete the parallelogram $ABCD$; and the diagonal AC will represent the resultant R; for it is the diagonal of a parallelogram whose sides represent the magnitude and direction of the components. Or lay off $AB = F$ units, $BC = F'$ units, making $ABC = 180° - \alpha$, the *supplement* of the angle made by the directions of the forces. The third side AC of the triangle thus formed, ABC represents R, since AB, BC, represent the components F, F'. (2.) *Trigonometrical Solution.*—Construct the triangle ABC as before. In it are given two sides, AB, BC, and the included angle $ABC = 180° - \alpha$, to find the third side AC.

76. Illustrations. To illustrate these methods let us take a very simple problem. Suppose a canal boat (Fig. 18) to be drawn by two horses, one on each bank, pulling in the directions AB, AC, by means of ropes attached to the bow of the boat, and making equal angles with the line of its keel. Let the pull exerted by each horse be 50 kgrs., the angle of inclination of their lines of action being $BAC = 90°$. It is required to find the resultant effect in moving the boat directly forward. To solve the problem graphically, lay off $AD = 50$ units, $AE = 50$ units. The angle $DAE = 90°$. Complete the parallelogram $ADFE$, and measure the number of units in the diagonal AF, which will be the number of pounds acting to move the boat along AG. The trigonometrical solution requires the determination of the hypothenuse AF of a right-angled triangle AEF, of which two sides, AE and $EF = AD$ are given. As each of these lines $= 50$, $AF = \sqrt{(50^2 + 50^2)} = 70 \cdot 71$ units, whence $R = 70 \cdot 71$ kgrs., or a little more than seven-tenths the sum of the components.

77. Composition of any number of Forces. Polygon of Forces. *If any number of forces are represented in magnitude and direction by all the sides but one of a polygon, the remaining side will represent a single equivalent force.*

This proposition follows from the *Parallelogram of Forces* in the same manner as the *Polygon of Motions* follows from the *Parallelogram of Motions.* Let AB, AE, AF, Fig. 19, represent three forces acting at a point A. The resultant of two of them, AB, AE, found by completing the parallelogram $ABCE$, is represented by the diagonal AC. Combining this resultant with the other force AF, by completing the parallelogram $ACDF$, we find AD to represent the magnitude and direction of the resultant of all three forces. But $AE = BC$, hence the components are also represented by AB, BC, CD, and these lines form three sides of a quadrilateral, of which AD is the fourth side. As this process of combination can be pursued indefinitely for any number of additional forces, the proposition is general.

78. Solution of Problems relative to the Composition of more than two Forces. Hence if a number of forces act simultaneously upon a body, their resultant may be found graphically, as follows: Construct a polygon by drawing lines proportional to the forces taken in order and parallel to their directions, and join the extremity of the line representing the last force with the starting point. The last side of the polygon thus formed will represent the resultant required.

For example, suppose a material point A, Fig. 20, to be acted upon by four forces, $AB = 5$, $AC = 10$, $AD = 15$. $AE = 20$, so situated that $BAC = 25°$, $CAD = 40°$, $DAE = 75°$. To find their resultant draw $ab = 5$ units of length, and parallel to AB; through b draw $bc = 10$, and parallel to AC; next draw $cd = 15$, and parallel to AD, and then $de = 20$, and parallel to AE. Finally join a and e; the length of line ae represents the magnitude of resultant, and the angle bae is the angle which its direction makes with the direction of the force AB.

The value of ae can also be determined by trigonometry by calculating successively ac from ab, bc and abc, ad from ac, ad and acd, and finally ae from ad, de and ade.

A simpler method of solution is furnished by analysis, which will be explained in the chapter treating of the analytical method of studying forces.

79. Parallelopiped of Forces. *If three forces not in the same plane are represented in magnitude and direction by three edges of a parallelopiped, the diagonal of that solid will represent a single equivalent force.*

At the point A suppose three forces applied, represented by the edges AE, AB, AD, of the parallelopiped AG, Fig. 21. Then will the diagonal AG represent their resultant. For by the *Parallelogram of Forces* the resultant of AB and AD is represented by AC. But AC is one side of a parallelogram $AEGC$, of which AE, adjacent to AC, is also a side. Hence the resultant of AC and the remaining force AE, which is the resultant of all three forces, is represented by AG, the diagonal of the parallelopiped.

The lines AB, $BC = AD$, $CG = AE$, form three sides of a *gauche* or twisted polygon (that is, a polygon whose sides are not all in the same plane), of which AG is the fourth side. Hence the present proposition is a particular case of the *Polygon of Forces*, and as the method of combination used could evidently be applied to additional forces lying in different planes, the *Polygon of Forces* is true, whether the lines of action of the forces lie in the same, or in different planes.

80. Equilibrium of Forces. Forces are said to be balanced, or in equilibrium, when their resultant is equal to zero, so that they counteract one another.

81. Equilibrium of two Forces. If two forces are applied at a single point, it is evident that *they will be in equilibrium only when equal and directly opposed.*

82. Equilibrium of three Forces. In the case of three forces in equilibrium, *the resultant of any two of them must be equal and opposite to the third.* For if the resultant of two of the forces is not equal and opposite to the remaining one, it may be combined with the latter, producing an unbalanced resultant force (§ 70), which is contrary to the supposition.

83. *If three forces acting upon a point can be represented in magnitude and direction by three sides of a triangle taken in order, they will be in equilibrium.*

At the point P, Fig. 22, let three forces, F, F', R, be applied in the directions PA, PB, PC, these forces being represented in magnitude and direction by the sides PA, AD, DP, of the triangle PAD *taken in order.* The resultant effect of PA and PB is a force represented by PD, which being the third side of the triangle PAD is equal and opposite PC. Hence the three forces are in equilibrium.

It is to be noted that the forces are represented by the sides *taken in order*; that is, the directions of the forces are those assumed by the sides in going round the triangle. The three forces at P act in the directions of lines drawn from P towards A, from A towards D, and from D towards P.

84. Conversely, *if three forces are in equilibrium about a point, they can be represented in magnitude and direction by three sides of a triangle drawn parallel to their lines of action.*

If the forces PA, PB, PC, are in equilibrium, any one of them, as PC, must be equal and opposite to the resultant PD of the other two. But PA, $AD = PB$, and PD, are three sides of the triangle PAD drawn parallel to the directions of the forces, hence PA, PB and PC, are represented by the three sides taken in order.

85. Since two triangles whose sides are perpendicular, each to each, are similar, it follows that *three forces in equilibrium about a point can be represented by three sides of a triangle drawn at right-angles to their lines of action.*

86. The theorem of the equilibrium of three forces may be put in still another form, which is sometimes useful in practice. From the demonstration of §. 82, putting the result in algebraic form, we have,

$$F : R :: PA : PD :: \sin PDA = \sin BPD : \sin PAD. \quad (12.)$$

$$F' : R :: PB : PD :: \sin PDB = \sin APD : \sin PBD. \quad (13.)$$

Also $BPD = 180° - BPC$, $APD = 180° - APC$, $PAD = 180° - APB$, $PBD = 180° - APB$.

Substituting these values in the preceding equations, we have,

$$F : R :: \sin BPC : \sin APB. \quad (14.)$$

$$F' : R :: \sin APC : \sin APB. \quad (15.)$$

Hence, *when three forces are in equilibrium, any one of them is proportional to the sine of the angle included between the directions of the other two.*

87. Equilibrium of any number of Forces. That any number of forces may be in equilibrium about a point, *the resultant of all but one of them must evidently be equal and opposite to the remaining force.*

88. *If the forces acting at a point can be represented in magnitude and direction by the sides of a polygon taken in order, they will be in equilibrium.*

For by a course of reasoning similar to that used in § 82, it will be seen that the resultant of all but one of the forces will be represented by the last side of the polygon, and hence will be equal and opposite to the remaining force.

89. Conversely, *if the forces acting at a point are in equilibrium, they can be represented in magnitude and direction by the sides of a polygon drawn parallel to their lines of action.*

The proof of this proposition is evidently precisely similar to that used in § 83.

90. Experimental Verification of Laws of Equilibrium. The laws of equilibrium of three forces may be shown experimentally by the apparatus represented in Fig. 23. Over two small pulleys, A, B, a cord is passed, to the ends of which weights are fastened. Weights, C, are also suspended from any point D on the loop, which is allowed to move freely, and assumes such a position that the three forces acting upon it are in equilibrium. If now we construct a triangle having its sides parallel to DA, DB, DC, respectively, we shall find that the lengths of those sides are proportional to the forces acting at D, parallel to their respective directions.

The triangle may be laid off upon paper, or on a blackboard placed behind the apparatus, or better still, we can make use of proportions (14), (15), measuring the angles ADC, BDC, ADB. Each of the three forces will be found to be proportional to the sine of the angle included between the direction of the other two. The angles are most conveniently measured by attaching a small bead to each of the threads (as shown in the figure), a decimetre distant from D. Then with a millimetre scale the distances of each bead from the others is measured. These distances will be the chords subtending the angles between the threads, which can then be obtained from a table of chords.[1]

If additional cords with weights attached to their extremities be fastened to D and passed over suitably-arranged pulleys, the apparatus of Fig. 25 becomes suitable for verifying the laws of the combination of a greater

[1] For a description of several new pieces of apparatus for class experiments relating to this subject, see a paper entitled *Apparatus Illustrating Mechanical Principles*, by R. H. Thurston; published in the *Journal of the Franklin Institute*, 3d Series, Vol. LXII., No. 3, p. 192. Also consult *Reference Table* of works on General Physics.

number of forces. It will be sufficient to allow D to assume its position of equilibrium, and to construct a polygon, having its sides parallel to the directions of the various forces. These sides will be found to have the same relative magnitude as the forces themselves. In the case of a number of combined forces, however, the friction of the pulleys and resistance of the cords causes a considerable variation between the theoretical and experimental results.

91. Practical Illustrations. The *Jib and Tie-Rod*, Fig. 24, employed in the common hoisting crane, furnishes a practical illustration of the preceding propositions. In the construction of such a machine, the frame must be made sufficiently strong to sustain the heaviest weight that is to be lifted, and in order to ascertain the strength to be given to each part, the maximum force to which it will be subjected must be known. This can be done by a simple application of the laws of equilibrium. Let P be the maximum downward pressure in kilogrammes, exerted upon the axle of the pulley at S (which can be determined when the maximum weight to be raised, and the angle WSD are known). This causes a certain pull (*strain*) upon the tie AB, in the direction BA, and a compression (*stress*) in the jib AC, in the direction AC. The rod AB will then be stretched, and the jib AC compressed, until the resistances to further change (which act in the directions AO, AM) produce a resultant equal and opposite to P. At this moment let the strain on AB be denoted by T, the stress in AC by T'. P, T and T' are in equilibrium about A, and hence may be represented by three sides of a triangle drawn parallel to their respective directions. From A draw AM, in the line of action of T', MN in that of T, and AN in the direction of P, that is, vertically downward. The forces P, T, T' are proportional to the three sides NA, MN, AM, of the triangle AMN, drawn parallel to their lines of action. That is,

$$P : T :: AN : MN :: BC : BA. \quad (16.)$$
$$P : T' :: AN : AM :: BC : CA. \quad (17.)$$
$$\text{whence } T = P\frac{BA}{BC}. \quad (18.) \qquad T' = P\frac{CA}{BC}. \quad (19.)$$

As P, BA, CA and BC are known quantities, the values of T, T' (expressed in kilogrammes) also become known, and these determined, the proper size of the beams AB, AC, can be calculated by means of the formulæ for the strength of beams.

Another interesting application of the foregoing principles is the mechanical contrivance known as the *Toggle-Joint*, or *Knee-Joint*, Fig. 25. It consists of two bars, AB, AC, connected by a joint at A, the other extremities resting upon the firm base PQ, and the movable plate MN, upon which a powerful pressure is to be exerted. A force P is applied at A by pulling or pushing in the direction AP. This evidently tends to raise MN, and thus presses upon any object confined between that plate and a fixed platform above it. The arms AB, AC, will evidently be compressed until the reactions thus developed are in equilibrium with P. To estimate their magnitude, upon AP lay off AE proportional to P. The reactions T, T', developed in AB, AC, will be represented by EF, FA, as AE, EF, FA, are three sides of a triangle drawn parallel to the directions of the forces in equilibrium. Hence

$$P : T :: AE : EF :: \sin AFE : \sin FAE. \quad (20.)$$
$$P : T' :: AE : FA :: \sin AFE : \sin FEA. \quad (21.)$$
$$\text{whence } T = P\frac{\sin FAE}{\sin AFE}. \quad (22.) \qquad T' = P\frac{\sin FEA}{\sin AFE}. \quad (23.)$$

An inspection of the figure will show that as MN rises under the influence of the force P, the angle FAD becomes more obtuse, whence also FAE and $DAE = FEA$ also increase, while AFE diminishes. Hence, under these circumstances the values of T, T', as given in (22), (23), also increase, becoming greater and greater as FAD approaches 180°, when T and T' equal infinity.[1] By the action of a comparatively small force we may thus produce an enormous pressure. The Toggle-Joint is frequently used in printing presses for bringing the type and paper into close contact, in machines for cutting large thicknesses of paper, in the cotton-presses at Mobile, etc. Its great advantage is that when in operation the pressure exerted by it increases simultaneously with the increase of resistance caused by the compression of the substance acted upon; it has the disadvantage, however, that to obtain a large range of vertical movement of the plate MN, the dimensions of the press must be very considerable.

92. The equilibrium of forces when the point of application is at rest, is known as *statical equilibrium*. When the forces acting upon a body in motion are balanced, we have *dynamical equilibrium*, the treatment of which is reserved for a future portion of this work.

93. Resolution of Forces. In paragraphs 70–78 we have explained the manner in which several forces can be replaced by a single resultant. We now take up the converse problem of finding two or more component forces, which are equivalent to a single force. This process is known as the *resolution of forces.*

Suppose AC, Fig 26, represent any force, and let it be required to find two other forces which acting together would produce the same effect as AC. It is merely necessary to construct any parallelogram $ABCD$ on AC as a diagonal. The sides AB, AD, will represent the required components. Or the triangle ADC may be constructed, in which case AD, DC, represent the components.

Since any number of triangles, ABC, ADC, AEC, AFC, Fig. 27, can be constructed on a single line taken as a base, it follows that if AC represent a force the lines AB and BC, AD and DC, AE and EC, AF and FC, will equally represent two component forces. Hence *any force can be resolved in an indefinite number of ways.*

To resolve a force into two components *whose directions are given*, we must know the angles BAC, BCA, between those directions and AC. We have then given two angles and the included side of a triangle, to find the remaining sides, which represent the components sought. This may be done either by trigonometry or graphically.

94. Resolution of a Force into two rectangular Components. It is frequently necessary to resolve a force into two components at right angles to each other. This is done by constructing a rectangle, or a right-angled triangle, upon a line

[1] Since in this case FAE and FEA each $= 90°$, and $AFE = 0°$, and $T = T' = P\frac{\sin 90°}{\sin 0°} = P\frac{1}{0} = \infty$.

representing the given force. Thus if AC, Fig. 28, represent this force, AB and AD, or DC and AD, will represent the components.

Calling R the original force, F the component represented by $A'B$ or DC, F' that represented by AD, and denoting the angle $BAC = ACD$ by α, in which case $DAC = 90° - \alpha$, we have $AB = AC \cos \alpha$, $AD = AC \sin \alpha$, whence $F = R \cos \alpha$ (24), $F' = R \sin \alpha$ (25), general equations for the resolution of a force into two rectangular components.

95. Equation of Relation between Components and Resultant. Denote the forces represented by AC, AB, BC, Fig. 27, by R, F, F', respectively, and call $BAC = \alpha$. $BCA = \beta$. Then as $AC = AB \cos BAM + BC \cos BCM$, $R = F \cos \alpha + F' \cos \beta$. (26.) Hence *the resultant of any two forces is equal to the algebraic sum of the products of each component, into the cosine of the angle which it makes with the resultant.*

96. Examples of Resolution of Forces. Let PQ, Fig. 29, be a body of weight W resting upon a horizontal table MN. It is required to find the downward pressure upon the table when PQ is acted upon by a force R, represented by CA. Let R be resolved into two rectangular components, one F, represented by CD, at right angles to MN, the other, F', represented by RB, parallel to MN. Of these forces, F alone exerts pressure upon the table, F' merely tending to make the body PQ slide along the surface on which it rests. Hence the whole downward pressure $= W + F$; or as $F = R \sin \alpha$, downward pressure $= W + R \sin \alpha$. The tendency to move horizontally is evidently $R \cos \alpha$.

Or, as another example, suppose that a body weighing W kilogrammes is to be raised by means of ropes AE, AF, Fig. 30, and it is wished to determine the number of kilogrammes which must be exerted at the end of each rope, in order to start it. Draw the vertical AC, having a length of W units, and complete the parallelogram $ABCD$ by drawing BC, CD, parallel to AF, AE. Resolve W into two components, one T, parallel to AE, and a second T' parallel to AF. Then T, T' will be the forces which must be applied at E and F to lift W. From the triangle of forces ABC we have

$$W : T :: AC : AB, \text{ whence } T = W\frac{AB}{AC} = W\frac{\sin DAC}{\sin CBA}. \quad (27.)$$

$$\text{and } W : T' :: AC : BC, \text{ whence } T' = W\frac{BC}{AC} = W\frac{\sin BAC}{\sin CBA}. \quad (28.)$$

from which T, T' becomes known if the angles made by each rope with the vertical are determined.

The examples given on p. 53 to explain the subject of statical equilibrium, also serve equally well to illustrate the resolution of forces.

97. Resolution of a Force into any number of Components. A force can also be directly resolved into any number of components by the principle of the *Polygon of Forces*, these components being in one or several planes. It is frequently simpler, however, to resolve the original force into two components, then one of these components into two others, and so on.

98. Practical Examples. As a good example of this, let us take the case of a vessel propelled by the wind. AB, Fig. 31, is a boat which is moved forward by the action of the wind blowing in the direction indicated by the arrow *W*. *MN* is the sail. To show how the vessel proceeds under an oblique wind, let *OP* represent the magnitude of the whole pressure of the wind upon the sail *MN*. Resolve it into two components, *OD* perpendicular to *MN*, and *OC* parallel to it. The parallel component *OC* can have no effect to move the vessel because it acts upon the sail edge-ways; hence the whole moving force is that due to the perpendicular component *OD*. Resolve *OD* into two other components, *OE* and *OF*, one of which, *OE*, is parallel to the keel of the vessel, and the other, *OF*, perpendicular to it. Then *OE* alone acts to push the vessel directly forwards, while *OF* tends to push it sideways. Hence the boat moves forward with a certain velocity due to the component *OE*, while at the same time it has a slight motion sideways (*leeway*), caused by the component *OF*. The reason why there is so little movement sideways in proportion to that forwards, is because of the greater resistance to motion in the former direction, caused by the shape of the vessel.

The manner in which two vessels can sail in different, and even opposite directions with the same wind, is easily demonstrated by the laws of the resolution of forces. In Fig. 32 the vessel *AB* is represented as proceeding in a direction opposite to that in which it moved in Fig. 31. The direction of the wind is the same as before, but the sail *MN* is placed in a new position. Resolving *OP* into *GD* and *OC* as before, *OC* is inoperative; *OD* can be again resolved into *OE* parallel to the keel, and *OF* perpendicular to it. Hence the vessel moves on in the line *BA* under the action of the component *OE*.

By varying the position of the sail, a vessel can be made to proceed in various directions, while the wind remains unchanged. In a boat with but one sail, the greatest advantage is gained when the wind is parallel to the keel, as its whole force is then exerted in causing a motion forward. The vessel is then said to be *scudding* or *sailing before the wind*. A ship sailing against the wind as closely as possible is said to be *close-hauled*. A large ship can sail so that her keel makes an angle of but six points (67° 30′) with the wind, and smaller vessels can sail very much closer. In a vessel with a large number of sails, a very favorable position of the wind is when it is at right angles to the keel (*upon the beam*), as the sails are then all acted upon with equal force, while if the wind is parallel to the keel the aft sails cut it off from those in front of them.

CHAPTER VI.

REACTION OF SURFACES. — COMPOSITION AND RESOLUTION OF FORCES ACTING AT DIFFERENT POINTS OF A BODY.

99. Equilibrium sustained by Reaction. When a body acted upon by a system of forces is at rest, equilibrium is often sustained by the reaction of one or more surfaces with which the body is in contact. The simplest example of this is when a body rests upon a horizontal plane, in which case the reaction caused by the compression of the material of which the plane is composed, is equal to the weight of the body. Additional examples will be found in the case of the *Jib and Tie-Rod* and the *Toggle Joint* already described.

100. Position of Resultant. If one surface be pressed against another in any manner, the resultant of the reactions at all the different points of its surface must be equal and opposite to the resultant of all the forces acting upon it.

101. Body pressed against a Curved Surface. When a body is pressed against a curved surface by the action of any number of forces, if there is equilibrium, the resultant of these must be normal to that surface. For if this were not the case, the resultant might be resolved into two other forces at right angles to each other, one normal to the curved surface, and the other tangential to it. The former would be opposed by the reaction of the surface, while the latter, being unopposed by any force, would produce motion over it, which is contrary to the supposition. The resultant must also pass through the point of contact of the surface, as otherwise there would be a tendency to rotate about that point. Thus let B, Fig. 33, be a cube pressed against the sphere A. For equilibrium, the resultant of the forces acting on A must pass through the point of contact P, and be perpendicular to the curved surface of A.

102. Constrained Bodies. If a body is in such a condition that motion can take place only in certain directions, it is said to be *constrained*. Thus a body fastened by a pivot is constrained to turn about that pivot. A body fastened at two points is constrained to move about an axis joining these two points, and if fastened at three points not in the same straight line, it is capable of no motion whatever.

103. Action of Forces on Constrained Bodies. It will be profitable to examine a few cases of the action of forces on constrained bodies.

The simplest case is that of a body resting upon an inclined plane. Let A, Fig. 34, be such a body, and let its weight, which acts vertically downward, be represented by the line W. Resolving this into two components, P perpendicular to the plane L, and F parallel to it, it is evident that the whole force of the component P is exerted in producing a pressure upon the plane at right angles to its surface, while the other component F tends to produce motion along L. A similar case of constrained motion will be noticed in the case of a body resting upon a horizontal surface, and acted upon by an oblique force.[1]

As another example, let P be a ring hung upon a curved wire AB, Fig. 35. If it be acted upon by a force F, normal to the curve, it will remain at rest, being kept in equilibrium by the reaction $-F$, equal and opposite to F. If, however, the force lie in any other direction, as in the line PR, it can be resolved into two components, one of which along PF is normal to the curve, and the other along PF' tangential to it. The former of these

[1] See p. 55.

will be balanced by the reaction of the wire, while the latter component will cause the ring to move along the wire.

When a body is fastened upon a pivot, any force F, Fig. 36, whose direction is in the line joining its point of application S with the pivot P, is directly opposed by the reaction of P. If the force has any other direction, as R, it will cause revolution about the pivot until the line PSR becomes straight, the component F producing pressure on P, while F' causes the body to rotate. F' evidently diminishes as the angle PSR increases, until it equals 0 when that angle assumes a value of $180°$.

104. Forces applied at different Points of a Body. Hitherto we have treated of forces applied at a single point. We now proceed to consider their action when applied at different points of any connected system of particles. The fundamental principle upon which our reasoning is based is the *Transferability of Force.*

The simplest case of transference is that of a force acting upon a rod in the direction of its length. The rod may be considered as composed of a line of particles, *abcd*, etc., Fig. 37. Now if a pull be exerted at A, the particle a is moved from its normal position, so that its distance from b is increased. This develops a molecular attractive force which acts upon b, causing it to move slightly towards a; thus in its turn c is acted upon, and so on through the whole length of the rod, until the molecular tension exerted between each particle is the same throughout the rod, and equal to the force applied at A. The rod is then in a state of equilibrium, and the particle n exerts a pull upon any point, as B, to which it is attached, equal to the force acting at A. Hence this power appears to be *transferred* from A to B. If, on the other hand, the rod is pushed at A in the direction AB, the distance between a and b is lessened, and the force thus developed is excited from particle to particle until the whole rod is in a state of tension. The particle n then exerts a force upon B equal to the pressure applied at A.

It follows from what precedes, that the effect upon B will be the same at whatever point of AB the power is applied. Hence *equal and opposite forces applied at the ends of a rod or rope, are in equilibrium.* For each may be supposed to be applied to any single particle taken in their line of action. Also the resultant of two or more forces thus applied must equal their algebraic sum.

105. Principle of Transferability of Force. *The efficiency of any force acting upon a body is not altered by transferring its point of application to any point in its line of action.* Thus let F be a force acting upon a body AB, Fig. 38. Evidently the effect of F is the same, whether it be applied at a, b, c, d or e.

This can be illustrated experimentally by balancing AB on a pivot P. The force F being applied at a, let it be counterpoised by a weight W, suspended from B. If then the point of application of F be changed from a to b, c, d or e, the weight will still be found to balance it exactly.

106. Composition and Resolution of Forces applied at different Points. The principle of transference furnishes a ready method of determining the resultant effect of two or more forces acting at different points of a body. Let F, F', be two forces applied to the body AB, Fig. 39, at the points M, N, and acting along the lines FM, $F'N$. Suppose these lines to be produced till they meet in some point S, which may lie either within or without the body AB. Then since S is in the continuation of FM, the effect of F will be the same as if applied at that point, and hence may be transferred to it. Also since S is in the continuation of $F'N$, F' may also be transferred in the same manner. The combined effect of the two forces is therefore the same as if both were applied at S, the intersection of their lines of action. In this case the resultant would be found, as already explained, by means of the parallelogram, so that if Sa, Sb represent F and F', SC will represent R, their resultant. Hence, to find the effect of two forces applied at different points of a body, *prolong their lines of action until they meet in a point, and proceed to find the resultant as if both forces acted at that point.*

The resultant of any number of forces applied at different points of a body may be found by combining them two by two.

107. Equilibrium of Forces applied at different Points. The resultant of F and F' would be balanced by the application of an equal and opposite force anywhere in the line SR. Hence, *if a body is kept at rest by the action of three oblique forces applied at different points, these forces would be in equilibrium if applied at a single point.*

108. Effect of Pivot. If a pivot be placed anywhere in the line of action of the resultant SR, as at C, the resultant may be considered as applied directly at that point, in which case it merely exerts a pressure upon the pivot without producing any tendency in the body to rotate around it. If the pivot does not lie in SR there will evidently be a tendency to rotation. Hence in the case of a body resting upon a pivot, *if the resultant of all the applied forces passes through the pivot, there will be equilibrium.* Conversely, *if the applied forces are in equilibrium the resultant passes through the pivot.*

CHAPTER VII.

Statical Moments.—Parallel Forces.—Couples.

Statical Moments.

109. Moment of a Force. *The tendency of a force to rotate a body about a fixed point is measured by the product of its intensity into the perpendicular distance from the point to the line of action of the force.* This product is called the *moment* of the force.

Thus the moment of the force F, Fig. 40, relatively to C is $F \times Cm$. Let F, F', Fig. 40, be two forces applied at M, N, and tending to cause rotation in opposite directions about a pivot placed at the point C, and let R be their resultant. When R passes through C there will be equilibrium among the forces F, F', and the reaction of the pivot, hence the tendencies of F, F' to rotate AB will in that case be equal. From C draw Cm, Cn, respectively, perpendicular to the directions of F, F', and construct a parallelogram $SbCa$, the adjacent sides of which, Sa, Sb, represent the magnitude and direction of F, F'. Then

$$F : F' :: Sa : Sb :: \sin bSC : \sin aSC :: \frac{Cn}{Sc} : \frac{Cm}{Sc} : Cn : Cm,$$

$$\text{whence } F \times Cm = F' \times Cn. \quad (27.)$$

But $F \times Cm$ is the moment of F, and $F' \times Cn$ the moment of F', relatively to C, and since we have shown that when their moments are equal the forces are balanced about C, these moments must represent the efficiency of the forces F, F', to cause rotation in either direction about that point.

Hence, calling F any force, l the perpendicular distance from any point to its line of action, and M its moment relatively to that point, we have

$$M = Fl. \quad (28.)$$

The perpendicular l is called the *arm* of the moment.

The direction of the tendency to rotation is said to be *right-handed* when it is in the direction of the moment of the hands of a watch, *i. e.*, from left to right, and *left-handed*, when in an opposite direction. Thus in Fig. 40, F tends to cause a right-handed, and F' a left-handed rotation. The former are generally designated by the sign $+$, the latter by the sign $-$.

110. Experimental Verification. The laws of moments may be verified experimentally by means of the apparatus represented in Fig. 41. AB is a disc of wood balanced on a pivot passing through C, so as to remain at rest indifferently in any position. A weight D is then attached at B by a cord, while a second cord is fastened to any other point of AB as K, and passed over a pulley M. Weights E are then attached to the latter cord, and the body AB is allowed to assume its position of equilibrium. There is then equilibrium between the moments of E and D. The perpendiculars CB, CL, dropped from C upon the directions of the forces, are then measured, and it will be found that $E \times CL = D \times BC$, in whatever position on the disc the point of attachment K be taken.

111. Resultant Moment of several Forces. If two or more forces tend to produce rotation in the same direction their united efficiency (*resultant moment*) is evidently equal to the sum of their moments.[1] If they tend to produce rotation in opposite directions, their united efficiency is equal to the excess of the sum of the moments in one direction over the sum of those in the opposite direction. Hence calling right-handed rotations +, and left-handed rotations —, *the resultant moment of any number of forces relatively to a point is equal to the algebraic sum of the moments of the components.* Or, calling the resultant moment M_r,

$$M_r = \Sigma Fl. \quad (29.)$$

The moment of the resultant of two forces acting to produce similar rotations is equal to the sum of the moments of the components, since the resultant may be substituted for them without change of efficiency; and the moment of the resultant of two forces acting to produce dissimilar rotations is equal to the difference of the moments of the components, for a like reason. Hence as this course of reasoning may be extended to any number of forces, it follows that *the moment of the resultant of any combination of forces equals the algebraic sum of the moments of the components.* Or, calling ΣFl this sum, R the resultant of all the forces, and l_0 the length of its arm,

$$Rl_0 = \Sigma Fl. \quad (30.)$$

As the preceding propositions hold, whatever may be the inclination of the forces to each other, they are true when the forces are parallel.

112. Equilibrium of Moments. When the sum of the moments in one direction equals the sum of the opposite moments, the resultant moment = 0, which is expressed algebraically,

$$Rl_0 = \Sigma Fl = 0. \quad (31.)$$

To produce this, R may become zero, or the arm l_0 may assume that value. If the latter is the case, there is a simple tendency to a translatory movement of the body along the line of action of the resultant. And if the body be fastened at any point in that line, the force acting upon it will be balanced by the reaction of the pivot or other support on which it rests, as already shown.

113. Moment of a Force relatively to an Axis. The moment of a force relatively to an axis is its tendency to produce rotation about that

[1] In this chapter we consider only the case in which all the forces lie in the same plane.

axis. Let R be a force represented by PR, Fig. 42, and AB an axis. Draw PM perpendicular to both PR and AB. If now the force PR be resolved into two components, one, F, represented by PF, perpendicular to AB, and the other F', represented by PF', parallel to AB, the former alone will have a tendency to produce rotation about that axis. The moment of the whole force R relatively to AB is therefore the same as the moment of its component F, which is evidently the moment of F relatively to the point M, that is to $F \times PM$.

Hence, *the moment of a force relatively to an axis is equal to the perpendicular distance between the axis and the line of action of the force, into that one of the components which is at right angles to the axis.*

Forces are in equilibrium about an axis when the algebraic sum of their moments relatively to it is zero.

114. Practical Application. An interesting application of the principle of equilibrium of forces about an axis is the following. If three equal weights, P, P, P, Fig. 43, be applied at equal distances from each other and from an axis through O, the algebraic sum of their moments relatively to that axis will always be equal to zero, whatever may be the absolute position of the points of application, A, B, C. The algebraic sum of the moments of P, P, P, is $P \times OH \times P + OL - P \times OD = P\ (OH + OL - OD)$, and it is to be proved that this product $= 0$. Join B, C, bisect BC in G and draw AG. As ABC is an equilateral triangle AG is perpendicular to BC and passes through O. Also $AO = 2OG$, whence $OD = 2OK$. Now $OH + OL = OH + OK + KL = 2OK$, as $HL = KL$ because of similarity of the triangles BKL, CKH. Hence $OH + OL - OD = OD - OD = 0$, and $P\ (OH + OL - OD) = 0$ for any position of A, B, C.

This is practically applied in pumping-machines in which three pumps are worked by a single shaft, their piston-rods being attached to three cranks, making angles of 120° with each other. The resistance of the three pumps is the same for all positions of the cranks, and perfect steadiness of motion is gained.

Parallel Forces.

115. Resultant of two Parallel Forces. *The resultant of two parallel forces acting in the same direction is parallel to them, and equal to their sum.*

Let F, F_1', Fig. 44, be any two forces applied at the points A, B. Their resultant, R, is found by prolonging FA, $F_1'B$, till they meet in P, and constructing a parallelogram of forces, $bPac$. It is clear that R may be considered as applied at P, and that its direction will lie between FA and $F_1'B$. Now imagine $F_1'B$ to be revolved about B in the direction indicated by the arrow. The point P continually recedes until the line $F_1'B$ assumes the position F_2B, parallel to FA, when P is at an infinite distance, and in this case, as all lines meeting at an infinite distance are parallel, PR, the direction of the resultant, must then be parallel to FA and $F_2'B$.

Also, for all inclinations of the components to each other, the magnitude of the resultant R of the forces F, F' is given by the equation $R = F \cos bPc + F' \cos aPc$ (26, p. 55). But when $F'B$ becomes parallel to FA, $bPc = 0$, $aPc = 0$, whence $R = F + F'$. (32.)

116. Point of Application of Resultant. To find the point of application of the resultant on the line AB, suppose M, Fig. 46, to be that point. Imagine a force R' equal and opposite to the resultant R to be applied there. This force would balance the resultant, and must therefore be in equilibrium with the components F, F'. Hence the moments of F, F' relatively to M, must be equal, that is, $F \times Ma = F' \times Mb$, whence $F : F' :: Mb : Ma$. But by similarity of triangles, $Mb : Ma :: MB : MA$, whence $F : F' :: MB : MA$. (33.) That is, *the resultant divides the line joining the points of application of the component forces into parts whose lengths are inversely proportional to the adjacent components.*

117. Two Parallel Forces in Opposite Directions. *If the forces are in opposite directions, the resultant is parallel to them, and equal to their difference. It lies on the same side of both, next the greater component, and has the same direction as that component.*

The three forces, F, F', R', Fig. 45, are in equilibrium, hence the resultant of F and R' must be equal and opposite to F'; that is, equal to $R' - F$ (Eq. 32). It is also evident from the figure that the resultant is on the same side of both F and R', next R', and in the same direction with it.

The point of application, B, is so situated that the moments of F, R' are equal, relatively to it, in which case $F : R' :: MB : AB$. (34.)

118. Practical Illustration. To illustrate the application of the preceding propositions, suppose two weights of 10 and 30 kgrs. respectively to be hung upon the extremities of a rod AB, Fig. 45, 2 metres in length. It is required to find the force R' which is requisite to support them, and the point at which it must be applied, neglecting the weight of the rod. We have $R' = R = F + F'$, and as $F = 10$ kgrs, $F' = 30$ kgrs., $F' = 40$ kgrs. Also $F : F' :: MB : MA$, or denoting AB by l and MB by x, $F : F' :: x$ $l - x$ whence $x = l \frac{F}{F + F'}$ (35) Substituting for l, F, F', their values as given above $x = 2 \times \frac{10}{40} = \frac{1}{2}$ m., and $l - x = 1\frac{1}{2}$ m. That is, the force R' must be applied at a distance of $\frac{1}{2}$ m. from M.

119. Resolution of a Force into two parallel Components. Proportions (33) (34) furnish a method of resolving a single force into two parallel forces applied at given points. Thus, let it be required to resolve the force R into two parallel forces applied at points A, B. In the proportion $F : F' :: MB : MA$ we know MB, MA, and in the equation $R' = R = F' + F'$, we know R. From these the value of F, F' can readily be computed. If F, F' are to have opposite directions, proportion (34), and the corresponding equation would be used in the same manner. It is also evident that if the magnitude of the components is given, their points of application can readily be found by the same proportions.

To illustrate, if the force $R' = 40$ kgrs. is to be resolved into two parallel components F, F' having the same direction, and applied at points A, B, distant from M, $1\frac{1}{2}$ m. and $\frac{1}{2}$ m. respectively, the magnitude of these is found as follows: $R = F + F' = 40$ kgr. $F : F' :: MB : MA :: \frac{1}{2} : 1\frac{1}{2}$ whence $F' = 3F$, and $R = F + F' = 4F = 40$ kgrs., whence $F = 10$ kgrs., $F' = 30$ kgrs.

120. Centre of Parallel Forces. Since the point of application M of the resultant of the parallel forces, F, F' divides the line AB joining the points of application of the forces in a fixed ratio depending only upon the magnitude of F, F', it follows that whatever may be the direction of the forces F, F', R, so long as their relative intensities remain unchanged, the position of M does not vary. This point is called the *Centre of the Parallel Forces.*

121. Resultant of any number of Parallel Forces. The position and magnitude of the resultant of more than two parallel forces may be found by combining them successively. Thus let F, F', F'', Fig. 46, be three parallel forces applied to a body at the points A, B, C. To find their resultant we first combine F and F'. Joining A and B we have from (33)

$$F : F' :: MB : MA,$$

which determines the point M. Also, $F = F + F'$. We next combine this resultant R with the remaining component F''. Joining MC, we have $R = F + F' : F'' :: CN : MN$, which determines the point of application N of the resultant R' of all three forces. Also $R' = R + F'' = F + F' + F''$. R' is evidently parallel to the components. The position of the centre, N, can be found either by a graphical construction, or calculated by the rules of trigonometry when the positions of A, B, C, are known.

If any one of the forces, as F'', acted in an opposite direction, we should evidently have $R' = F + F' - F''$, and its position would be found by (34).

As this process of combination can be continued indefinitely, it follows that calling forces acting in one direction $+$, those in an opposite direction $-$, *the resultant of any number of parallel forces equals their algebraic sum*, that is,

$$R = \Sigma F. \quad (36.)$$

The preceding method of determining the position of the point of application of the resultant is simple when there are but few forces to be considered, but becomes very tedious with numerous forces. Another method is then adopted, which will be explained in the following chapter.

122. Experimental Demonstration. The laws of parallel forces can be demonstrated experimentally by means of a very simple apparatus represented in Fig. 47. A graduated bar, AB, is suspended from the hooks of two delicate spring balances D, E. The weight of AB causes a certain constant depression of the indexes of the balances, which is read once for all. A weight W is then placed anywhere upon AB, as at C, and the balance readings taken anew, which being diminished by the preceding readings, show the pressures exerted by it at G and H, the points of suspension. Calling these F, F', it will be found that in whatever position W may be placed, $W = F + F'$, and $F : F' :: CH : CG$. For verifying the laws with a greater number of parallel forces additional balances may be placed between D and E.

Couples.

123. Definition. If two equal and opposite parallel forces not acting in the same straight line be applied to a body, their algebraic sum is 0, and hence they have no tendency to produce a motion of translation. But as their resultant moment with regard to any point can never equal 0, their effect will be to rotate the body about an axis.

Such a combination of forces is called a *couple*. The perpendicular distance between the lines of action of the forces is called the *arm of the couple*, and the plane containing these lines of action the *plane of the couple*. Any line at right angles to this plane is an *axis of the couple*. The terms *right-handed* and *left-handed* are applied to couples in the same manner as in the case of moments.

124. Efficiency of a Couple. *The rotary effect, or moment of a couple is measured by the product of either force into the length of the arm.*

Let F, F', Fig. 48, constitute a couple whose arm is AB. Assume an axis at right angles to the plane of F, F', passing through any point P. The moment of F relatively to $P = F \times PB$, and the moment of F' relatively to the same point $= F' \times PA = F \times PA$, as $F' = F$. The resultant moment of the two forces is therefore equal to

$$F \times PB + F \times PA = F(PB + PA) = F \times AB.$$

If the axis be taken at a point not between the lines of action of F and F', as P', we have *moment of force F relatively to $P' = F \times P'B$, moment of force F relatively to $P' = -F' \times P'A = -F \times P'A$*, as the moment is left-handed. Hence, *resultant moment of F and $F' = F \times P'B - F \times P'A = F(P'B - P'A) = F \times AB$*, as before.

Hence calling M the moment of the couple, F the magnitude of the force acting at each end, and l the arm, $M = Fl$. (37.)

125. Transference of Couples. *A couple may be turned in its own plane, or moved parallel to itself without altering its efficiency.*

For the effect of the couple F, F', Fig. 49, to cause rotation about P, is evidently not altered if the forces assume the positions f, f', as the arm MN equals AB. With any axis, as P', not situated between the lines of action of F, F', we have *Rotary-effect* $= f \times mn = f \times MN = F \times AB$. The moving of the couple to any other position in the same plane would simply be equivalent to moving the assumed axis to some other point, as from P to P', which does not alter the efficiency.

A couple may also be transferred to any plane parallel to its own plane without alteration of its efficiency, because its whole tendency being to produce rotation about an axis at right angles to its plane, the effect will be the same in whatever plane at right angles to the axis it may lie.

126. Reduction of Couples. From Eq. (37) it follows that couples are equivalent when their moments are equal. That is, a couple of force $= 8$ and arm $= 3$, has the same efficiency as one of force $= 2$ and arm $= 12$, as in either case $M = Fl = 24$. Hence *any couple may be replaced by an equivalent one having a given arm.* The value of the force in this case is found from the equation $M = Fl$, whence $F = \frac{M}{l}$ (38). If, on the other hand, we wish the new couple to have a given force, we may find the corresponding length of the arm from the equation $l = \frac{M}{F}$ (39). This process is called the reduction of couples to the same arm, or the same forces. For example, let it be required to replace the couple $F = 5$, $l = 10$, by an equivalent one having an arm $= 2$. From (38) we find the corresponding value of F to be $\frac{M}{l} = \frac{50}{2} = 25$. An equivalent couple having an arm 2 must then have a force of 25.

127. Equilibrium of Couples. Since the sole effect of a couple is to produce rotation, no single force can be so applied as to hold it in equilibrium. It will evidently be balanced, however, by the application of an equal couple acting in an opposite direction.

128. Combination of Couples having the same Axis. *The resultant moment of any number of couples lying in the same or parallel planes, is equal to the algebraic sum of their separate moments.*

For they may all be reduced to equal arms, and so applied that the forces act in the same straight line. Calling Fl, $F'l$, $F''l$, $-F'''l$, the reduced couples, these will be equivalent to a single couple with a force $= F + F' + F'' - F'''$, and an arm l, whose efficiency equals the algebraic sum of the component couples.

Hence denoting the resultant moment of any number of couples by M_r, we have $M_r = \Sigma Fl$. (40.)

For equilibrium we must have $\Sigma Fl = 0$. (41.)

129. Representation of Couples by Lines. In many cases, especially where the planes of the couples are inclined to each other, it is convenient to represent the direction, magnitude and intensity of a couple in the following manner. Let F, F', Fig. 48, be the couple which is to be represented. From any point O draw a line OM at right angles to the plane in such a direction that to an observer looking from O towards M the couple shall seem right-handed, and let the length of OM be made proportional to the number of units in the moment of the couple.

130. Combination of Couples having different Axes. Couples lying in planes inclined to each other can be combined by means of a theorem similar to the *Parallelogram of Forces*. Let F, F', f, f', Fig. 50, be two couples reduced to the same arm AB and acting in planes FAF' faf', making with each other any angle FAf. Let the lines FA, fA, $F'B$, $f'B$ represent the magnitudes of the forces of the couple acting at the extremities of AB. The two forces F, f, give a resultant R represented in magnitude and direction by AR; and F' f' give an equal and parallel resultant R', represented by BR'. The four forces F, f, F' f' may therefore be replaced by two parallel forces R, R', lying in the plane RAB. The resultant moment evidently equals $R \times AB$. Hence couples may be combined and resolved in the same manner as ordinary forces. If instead of the forces F, F' f, f' we represent by AF, AF', Af, Af' the *moments* of the couples, AR, AR', will evidently represent the moment of the resultant couple.

Since the plane of a couple and its axis of rotation are at right angles the axes have the same inclinations to each other as the planes. Hence the moments of F, f may be laid off on the axes AF, Af, Fig. 51, instead of in the planes BC, BE. The line AR will then represent the moment of the resultant couple, the plane of which is BD, at right angles with AR. We have, therefore, the general proposition, *if the moments and axes of two couples be represented by two sides of a parallelogram, the diagonal of that parallelogram represents the moment and axis of the resultant couple.*

131. Illustrations. The effect of two equal and opposite forces to cause rotation is seen in spinning a common humming-top, the pull exerted on the string and applied at the circumference of the axle being one of the forces, and the reaction of the handle applied at the axis of symmetry of the axle the other. Another excellent example is the case of a light sphere of cork or wood kept rapidly rotating in the air by the action of a fountain-jet playing against the under side of it. The upward force of the jet applied on the surface is equal and opposite to the force of gravity (or weight of the ball) applied at the centre of the sphere, thus forming a couple.

CHAPTER VIII.

ANALYTICAL STATICS.

132. General Principles, Definitions, etc. When we have a large number of forces to consider, the processes of composition and resolution of forces are very much simplified by the use of analytical methods.

It is customary in this case to refer the directions of the forces, their points of application, etc., to coördinate axes at right angles to each other, as in Analytical Geometry.

To the student unacquainted with that study the following explanation will give the main principles required for an intelligent study of the present chapter.

When the forces are all in the same plane we make use of two rectangular axes OX, OY, Fig. 52, known as the *axis of* X and *axis of* Y respectively, and collectively as the *axes of coördinates*. Lines drawn from any point parallel to the axis of Y and terminated by the axis of X are called *ordinates*, and are generally designated by the letter y, while lines drawn from a point parallel to the axis of X and terminated by the axis of Y are called *abscissas*, and are designated by the letter x. The abscissas and ordinates of a point taken together are called the *coördinates* of that point. Thus BM and $BN = MO$ are coördinates of B, BM being the ordinate, BN the abscissa. The point O at the intersection of the axes is the *origin of coördinates*. Abscissas of points lying to the right of the axis of Y are generally affected by the sign $+$; of those lying to the left, $-$. Ordinates of points lying above the axis of X are $+$; of points below, $-$. Evidently the values of x and y for any point fix its position; thus the position of B is known when we have given $OM = x$, $MB = y$. To illustrate the application of this to mechanics; two forces of magnitude 5 and 4 applied at a single point, and making angles of 45° and 30° with any given line are represented by taking that line as the axis of X (or Y if preferable), and the point of application as O, and laying off two lines OA, OB of lengths $=$ 5 and 4 units respectively, making angles of 45° and 30° with OX.

When the forces do not lie in a single plane they can be referred to three rectangular axes, X, Y, Z, Fig. 54, formed by the intersection of three planes at right angles to each other. The position of a force in space is then represented by the angles α, β, γ, which its direction makes with the axes of X, Y and Z. The coördinates of any point referred to a system of three coördinate axes are denoted by the letters x, y, z, indicating distances parallel to X, Y or Z respectively.

The *projection* of a line upon a plane is the distance between perpendiculars drawn to the plane from each of its extremities. The projection of a line upon another line is the distance between perpendiculars drawn to the latter line from the extremities of the former. The projection of a line upon a plane or a second line is equal to the length of the line into the cosine of angle which it makes with the plane or line upon which it is projected.

133. Analytical Statics in Two Dimensions.—Forces applied at a single Point. Let F, represented by OB, Fig. 52, be any force applied at a single point. It can be resolved into two rectangular components represented by MO and BM. Denote the angle BOM by α.

Then, as $MO = BO \cos \alpha$, $BM = BO \sin \alpha$, it is clear that the component parallel to $X = F \cos \alpha$, and that parallel to $Y = F \sin \alpha$.

The direction of the components will be found by observing the algebraic signs of $\cos \alpha$ and $\sin \alpha$. Thus if the force F lie in the direction indicated by the line OC, the angle α which it makes with X is MOC (counted on the same side of OX with OB), and both $\cos \alpha$ and $\sin \alpha$ are *minus*, MOC being greater than 180°. In this case the components would be $-F \cos \alpha$, parallel to X, $-F \sin \alpha$ parallel to Y.

14. Resultant of any Number of Forces. Suppose that any number of forces. F, F', F'', F''', F^n, are applied at a single point, and it is required to find their resultant. Take the point of application as the origin of coördinates, and denote by α, α', α'', α''', α^n, the angles made by the directions of the forces with the axis of X. Denote the resultant by R, and call θ the angle which it makes with X. Resolving F, F', etc., into their components parallel to X and to Y, as explained in the preceding paragraph, and taking care to observe the algebraic signs of the functions $\sin \alpha$, $\cos \alpha$, etc., the results of the following tables are obtained.

Components parallel to X.	*Components parallel to Y.*
$F \cos \alpha$	$F \sin \alpha$
$F' \cos \alpha'$	$F' \sin \alpha'$
$F'' \cos \alpha''$	$F'' \sin \alpha''$
$F''' \cos \alpha'''$	$F''' \sin \alpha'''$
$F^n \cos \alpha^n$	$F^n \sin \alpha^n$

Hence, *Total force along*

$$X = F \cos \alpha + F' \cos \alpha' + F'' \cos \alpha'' + F''' \cos \alpha''' + F^n \cos \alpha^n = \Sigma F \cos \alpha.$$

Total force along

$$Y = F \sin \alpha + F' \sin \alpha' + F'' \sin \alpha'' + F''' \sin \alpha''' + F^n \sin \alpha^n = \Sigma F \sin \alpha.$$

Now conceive the resultant R to be resolved in the same manner. Its components are

$R \cos \theta$, *along* X, $\qquad$ $R \sin \theta$, *along* Y.

But the components of R along X must evidently be equal to the sum of the components of the elementary forces F, F', etc., along the same axis, and the components of R along Y must be equal to the sum of the components of F, F', etc., along that axis. Hence

$$R \cos \theta = \Sigma F \cos \alpha. \quad (42.) \qquad R \sin \theta = \Sigma F \sin \alpha. \quad (43.)$$

Squaring these equations and adding the results,

$$R^2 \cos^2 \alpha + R^2 \sin^2 \alpha = (\Sigma F \cos \alpha)^2 + (\Sigma F \sin \alpha)^2, \text{ or as } \cos^2 \alpha + \sin^2 \alpha = 1,$$

$$R^2 = (\Sigma F \cos \alpha)^2 + (\Sigma F \sin \alpha)^2; \text{ whence}$$

$$R = \sqrt{(\Sigma F \cos a)^2 + (\Sigma F \sin \alpha)^2}. \quad (44.)$$

Equation (44) gives the magnitude of the resultant. It remains only to determine its direction by means of the angle θ.

Dividing Eq. (43) by Eq. (42), $\dfrac{R \sin \alpha}{R \cos a} = \dfrac{\Sigma F \sin \alpha}{\Sigma F \cos \alpha}$, whence

$$\text{tang } \theta = \frac{\Sigma F \sin \alpha}{\Sigma F \cos \alpha}. \quad (45.)$$

The algebraic sign of tang θ evidently follows from the signs of $\Sigma F \sin \alpha$, $\Sigma F \cos \alpha$.

The value of R may also be obtained (having first determined the angle θ) from the equation $R \cos \theta = \Sigma F \cos \alpha$, whence

$$R = \frac{\Sigma F \cos \alpha}{\cos \theta} = (\Sigma F \cos \alpha) \sec \theta. \quad (46.)$$

That there may be equilibrium among a system of forces R must $= 0$, whence $(\Sigma F \cos \alpha)^2 + (\Sigma F \sin \alpha)^2 = 0$, (47), which equation can only be satisfied when each of its terms $= 0$. Hence the equations of equilibrium are

$$(\Sigma F \cos \alpha) = 0, \quad (48.)$$
$$(\Sigma F \sin \alpha) = 0. \quad (49.)$$

135. Resultant of Oblique Forces applied at several Points. In this case there is a tendency to rotation about an axis produced as well as a resultant translatory force.

Let F, F', F'', F^n be forces applied at points whose coördinates relative to an assumed set of coördinate axes X, Y, Fig. 53, are (x, y), (x', y'), (x'', y''), (x^n, y^n). It is required to find the magnitude and direction of the resultant, and also the resultant couples produced. Their directions are denoted by the angles α, α', α'', α^n, as before.

Let us first consider the case of a single force F applied at the point S (x, y).

Resolving it into two components we have

$$P = F \cos \alpha = \textit{component parallel to } X.$$
$$Q = F \sin \alpha = \textit{component parallel to } Y.$$

Now let us imagine two opposite forces K, $-K$, each equal and parallel to P, to be applied at O. Also two opposite forces T, $-T$, each equal and parallel to Q. The action of the system of forces F, F', etc., will not be altered by this addition. We have now acting upon the body, (1) at the point S (x, y), the forces $P = F \cos \alpha$, parallel to X, and $Q = F \sin \alpha$ parallel to Y; (2) at the point O, the forces $+K$, $-K$, along X, and $+T$, $-T$, along Y.

The combined effect of these forces is as follows; — the forces K and T are together equal to a single force

$$L = \sqrt{K^2 + T^2} = \sqrt{(F \cos \alpha)^2 + (F \sin \alpha)^2} = F,$$

making an angle β with X of such value that

$$\text{tang } \beta = \frac{T}{K} = \frac{Q}{P} = \frac{F \sin \alpha}{F \cos \alpha} = \text{tang } \alpha, \text{ whence } \beta = \alpha,$$

that is, L is equal and parallel to F and applied at O.

The equal and opposite forces P and $-K$ form a right-handed couple, having an arm $SM = y$, and whose moment is consequently $Py = yF \cos \alpha$. Also the equal and opposite forces Q and $-T$ form a left-handed couple with an arm $SN = x$, and whose moment is $-Qx = -xF \sin \alpha$. These couples being in the same plane, form a resultant couple

$$M = Py - Qx = yF \cos \alpha - xF \sin \alpha.$$

Now suppose this process to be repeated for each of the forces F', F'', F^n. We shall then have a similar result for each: that is, the force F' will be replaced by an equal and parallel force applied at O, and a couple whose moment $= y'F' \cos \alpha' - x'F' \sin \alpha'$; the force F^n will be replaced by an equal and parallel force applied at O, and a couple $y^nF^n \cos \alpha^n - x^nF^n \sin \alpha^n$, and so for all the forces. Their combined efficiency will therefore be the resultant of F, F', F'', F^n applied at O, and the resultant of all the couples, which has a moment equal to the algebraic sum of the elementary couples. Using the same notation for the resultant as in the preceding cases, we have as the joint effect of all the forces,

$$R = \sqrt{(\Sigma F \cos \alpha)^2 + (\Sigma F \sin \alpha)^2}, \text{ applied at } O, \quad (50.)$$

$$\text{tang } \theta = \frac{\Sigma F \cos \alpha}{\Sigma F \sin \alpha}, \quad (51.)$$

and the couple of moment $= \Sigma\, (yF \cos \alpha - x\, F \sin \alpha)$. (52.)

If $\Sigma\,(yF\cos\alpha - xF\sin\alpha) = 0$, the resultant passes through O. If $R = 0$ and $\Sigma\,(yF\cos\alpha - xF\sin\alpha)$ does not $= 0$, the resultant is simply a couple.

When there is equilibrium among the forces applied at different points of a body

$$R = 0, \text{ whence } (\Sigma F\cos\alpha) = 0,\ (\Sigma F\sin\alpha) = 0, \quad (53.)$$

$$\text{and } \Sigma\,(yF\cos\alpha - xF\sin\alpha) = 0. \quad (54.)$$

136. Analytical Statics in three Dimensions.—Resolution of a Single Force into three Rectangular Components. Let F, Fig. 54, be any force in space which is required to be resolved into three components at right angles to each other. Take the point of application of F as the origin of coördinates, and let OF represent that force, its direction being indicated by the values of α, β, γ, the angles which its line of action makes with the axes X, Y, Z respectively. Construct the rectangular parallelopiped $OBFE$ having OF as its diagonal. The adjacent edges OA, OB, OC will represent the components required (§ 79, p. 50). It only remains to find analytical expressions for these in terms of F.

Project OF upon the axes OX, OY, OZ, successively. The figures $ODFA$, $OEFB$, $OCFG$, all being rectangles,

$$OA = OF\cos\alpha,\ OB = OF\cos\beta,\ OC = OF\cos\gamma.$$

But OA, OB, OC represent the rectangular components of F, hence denoting these by P, Q, S, we have

$$P = F\cos\alpha \ (53). \quad Q = F\cos\beta \ (54). \quad S = F\cos\gamma \ (55.)$$

137. Composition of Three Forces at Right Angles. Conversely, the resultant of three rectangular components can be found analytically as follows:—

Squaring equations (53), (54), (55), and adding the resultant we have

$$P^2 + Q^2 + S^2 = F^2\cos^2\alpha + F^2\cos^2\beta + F^2\cos^2\gamma =$$

$$F^2(\cos^2\alpha + \cos^2\beta + \cos^2\gamma).$$

But as $\cos^2\alpha + \cos^2\beta + \cos^2\gamma = 1$ (by a proposition of analytical geometry), $F^2 = P^2 + Q^2 + R^2$, and $F = \sqrt{P^2 + Q^2 + R^2}$. (56.)

The same result may be reached geometrically, as

$$OF^2 = OA^2 + OB^2 + OC^2. \quad (57.)$$

138. Resultant of any Number of Forces in Space applied at a Single Point. Let F, F', F'', F^n be any forces applied at a single point. Take their point of application as the origin of coördinates, and denote the angles made by the forces, and the axes of coördinates by the letters α, β, γ, α', β', γ', α'', β'', γ'', α^n, β^n, γ^n respectively. Denote their resultant by R, and let its direction be indicated by the angles α^r, β^r, γ^r, which it makes with X, Y, Z.

Now resolve F, F', etc., into rectangular components along X, Y, Z, as explained in § 136, taking care to observe the algebraic signs of the cosines of the direction-angles. These components are

Along X.	*Along Y.*	*Along Z.*
$F\cos\alpha$	$F\cos\beta$	$F\cos\gamma$
$F'\cos\alpha'$	$F'\cos\beta'$	$F'\cos\gamma'$
$F''\cos\alpha''$	$F''\cos\beta''$	$F''\cos\gamma''$
$F^n\cos\alpha^n$	$F^n\cos\beta^n$	$F^n\cos\gamma^n$

From these it follows that

Total force along $X =$

$$F\cos\alpha + F'\cos\alpha' + F''\cos\alpha'' + F^n\cos\alpha^n = \Sigma F\cos\alpha.$$

Total force along $Y =$

$$F\cos\beta + F'\cos\beta' + F''\cos\beta'' + F^n\cos\beta^n = \Sigma F\cos\beta.$$

Total force along $Z =$

$$F\cos\gamma + F'\cos\gamma' + F''\cos\gamma'' + F^n\cos\gamma^n = \Sigma F\cos\gamma.$$

Now conceive the resultant R to be resolved into three components along the axes of coördinates. These will be

$$R \cos \alpha^r \text{ along } X.$$
$$R \cos \beta^r \text{ along } Y.$$
$$R \cos \gamma^r \text{ along } Z.$$

But the rectangular components of R along X, Y, Z respectively must evidently be equal to the sum of the components F, F', F'', F^n along the same axes. Hence

$$R \cos \alpha^r = \Sigma F \cos \alpha, \quad (58.)$$
$$R \cos \beta^r = \Sigma F \cos \beta, \quad (59.)$$
$$R \cos \gamma^r = \Sigma F \cos \gamma. \quad (60.)$$

Squaring (58), (59), (60) and adding the results,

$$R^2 (\cos^2 \alpha^r + \cos^2 \beta^r + \cos^2 \gamma^r = (\Sigma F \cos \alpha)^2 + (\Sigma F \cos \beta)^2 + (\Sigma F \cos \gamma)^2,$$

$$\text{or } R = \sqrt{(\Sigma F \cos \alpha)^2 + (\Sigma F \cos \beta)^2 + (\Sigma F \cos \gamma)^2}. \quad (61.)$$

This same value of R may also be found by considering R as a single force with components $\Sigma F \cos \alpha$, $\Sigma F \cos \beta$, $\Sigma F \cos \gamma$, and using the same geometrical reasoning as in the case of determining any force from its rectangular components as already explained (§ 137).

The direction of R is given by the equations

$$\cos \alpha^r = \frac{\Sigma F \cos \alpha}{R}, \ (62.) \quad \cos \beta^r = \frac{\Sigma F \cos \beta}{R}, \ (63.) \quad \cos \gamma^r = \frac{\Sigma F \cos \gamma}{R}. \ (64.)$$

For equilibrium,

$$R = \sqrt{(\Sigma F \cos \alpha)^2 + (\Sigma F \cos \beta)^2 + (\Sigma F \cos \gamma)^2} = 0, \quad (65.)$$

in which case

$$(\Sigma F \cos \alpha)^2 = 0, \quad (\Sigma F \cos \beta)^2 = 0, \quad (\Sigma F \cos \gamma)^2 = 0. \quad (66.)$$

Hence equations (66) are the equations of equilibrium.

When forces in space are applied at different points they tend to produce a translatory motion, and also rotation about each of the three axes X, Y, Z. This case being quite complex, the student is referred for its demonstration to more extended works on mechanics.

139. Resultant and Centre of any Number of Parallel Forces in Space. A much simpler method of obtaining the position of the centre of parallel forces than the construction explained in § 121, p. 64 is afforded by analytical methods based on the principle of moments.

Let R be a force applied at the point S whose coördinates are x^n, y^n, z^d (Fig. 55). If R be parallel to OY its moment relatively to the axis OZ is Rx. If parallel to OZ its moment relatively to OX is Ry, and if parallel to OX its moment relatively to OY is Rz.

Now suppose any number of parallel forces F, F', F^n applied at points (x, y, z), (x, y', z'), (x^n, y^n, z^n) to act upon a body, and suppose S, Fig. 56, to be the point of application of their resultant R (centre of parallel forces). The magnitude of the resultant is equal to the sum of the components or, $R = \Sigma F$. The coördinates of S are determined as follows: — Since the parallel forces composing the system may be turned in any direction without altering the position of the centre (§ 120, p. 64), we may consider them as acting parallel to each of the axes successively. First let us suppose them to be parallel to OY. The moment of the force F relatively to OZ is then Fx; that of F', $F'x'$; that of F^n, $F^n x^n$. The sum of these moments is ΣFx, and this must be equal to the moment of the resultant, which is $x^r \Sigma F$. Hence $x^r \Sigma F = \Sigma Fx$. Next suppose all the forces to be parallel to OZ. By the same course of reasoning we have $y^r \Sigma F = \Sigma F y$. Then supposing all the forces to be parallel to OX, we have $z^r\ \Sigma F = \Sigma F z$ From these three equations

$$x^r\ \Sigma F = \Sigma Fx, \quad (67.)$$
$$y^r\ \Sigma F = \Sigma Fy, \quad (68.)$$
$$z^r\ \Sigma F = \Sigma Fz, \quad (69.)$$

we find the required coördinates to be

$$x^r = \frac{\Sigma Fx}{\Sigma F}, \quad (70.) \quad y^r = \frac{\Sigma Fy}{\Sigma F}, \quad (71.) \quad z^r = \frac{\Sigma Fz}{\Sigma F}, \quad (72.)$$

For equilibrium there must be no tendency either to translation or rotation, hence $R = \Sigma F = 0$, (73.) $\Sigma F\,x = 0$, $\Sigma F\,y = 0$, $\Sigma F\,z = 0$, (74.) are the equations of equilibrium.

140. Virtual Velocities.—Definition. Let F, Fig. 56, be a force applied at the point A, and suppose this point of application to be moved by some external force through an extremely small space, so that it assumes the position A' or A'', the force F still acting in the same direction as at first. From A' or A'' draw a perpendicular $A'B$ or $A''B'$. The distance AB or AB' intercepted between the original point of application A and the foot of this perpendicular, is called the *virtual velocity* of the force F, and is positive when it is laid off in the direction in which the force acts, and negative when opposite to that direction. Thus AB is positive, AB' negative.

141. Principle of Virtual Velocities. *If the forces acting upon a body are in equilibrium, and a very small displacement be given to the body, the algebraic sum of the products of each force into its virtual velocity is zero.*

If the forces are applied at a single point, represent this point by A, Fig. 57, and the forces themselves by F, F', F'', F^n. Suppose the point of application to be moved over the very small distance AA'. Call α, α', α'', α^n the angles made with X by the directions of the forces, β, β', β'', β^n the angles made with Y, and denote the angle between X and the line AA' by θ. AB is the virtual velocity of F, which may be called v. Denote the virtual velocities of the other forces by v', v'', v^n respectively. It is to be proved that $\Sigma Fv = 0$.

For any single force, as F,

$$v = AB = AA' \cos BAA' = AA' \cos(\alpha - \theta) = AA' (\cos\alpha\cos\theta + \sin\alpha\sin\theta),$$

$$\text{whence } Fv = F \times AA' (\cos\alpha\cos\theta + \sin\alpha\sin\theta). \quad (75.)$$

In like manner for F', F'', F^n we have

$$F'v' = F' \times AA' (\cos\alpha'\cos\theta + \sin\alpha'\sin\theta), \quad (76.)$$
$$F''v'' = F'' \times AA' (\cos\alpha''\cos\theta + \sin\alpha''\sin\theta), \quad (77.)$$
$$F^n v^n = F^n \times AA' (\cos\alpha^n\cos\theta + \sin\alpha^n\sin\theta), \quad (78.)$$

Taking the sum of these equations, noting that α, β, α', β', etc., are complements, so that $\sin\alpha = \cos\beta$, $\sin\alpha' = \cos\beta'$, etc.,

$$Fv + F'v' + F''v'' + F^nv^n = \Sigma Fv = AA' [\cos\theta\,(F\cos\alpha + F'\cos\alpha' + F''\cos\alpha'' + F^n\cos\alpha^n) + \sin\theta\,(F\cos\beta + F'\cos\beta' + F''\cos\beta'' + F^n\cos\beta^n)]. \quad (79.)$$

But, as by supposition, there is equilibrium of the forces about A, it follows from equations (48), (49), p. 70, that

$$\Sigma F\cos\alpha = F\cos\alpha + F'\cos\alpha' + F''\cos\alpha'' + F^n\cos\alpha^n = 0,$$
$$\Sigma F\cos\beta = F\cos\beta + F'\cos\beta' + F''\cos\beta'' + F^n\cos\beta^n = 0,$$

and as the terms of equation (79) enclosed between the brackets become $= 0$, the whole member $= 0$, and hence $\Sigma Fv = 0$. (80.)

It follows from (80) that the resultant of any number of forces multiplied by its virtual velocity equals the algebraic sum of each of its components multiplied into the corresponding virtual velocity, that is, if we call V the virtual velocity of the resultant R, $RV = \Sigma Fv$. (81.)

In case the forces are not applied at a single point, the theorem still holds. For suppose the forces F, F', F'', F^n to be applied at different points. It follows from the principle explained in § 106 (p. 59) that any two of them, as F, F', produce the same effect as if they were both applied at the intersection of their lines of action. Imagine them thus applied, call R the resultant equal to their combined effect, and denote by m, n, the virtual velocities of F, F', when transferred to the point supposed. According to the principles stated in the preceding paragraph,

$$RV = Fm + F'n. \quad (82.)$$

Now the virtual velocities of F, F' are not affected by the transference of their points of application. For suppose A, D, Fig. 58, to be these points, and let the body be slightly displaced by rotation, though a very small angle a about an axis passing through any point M, and at right angles to the plane of the forces. The points A, D, move through very small arcs AC, DF, which may be considered as straight lines at right angles to the radii MA, MD. Also $AC = MAa$, $DF = MDa$. AB is the virtual velocity of F, DE of F'. If F, F' were transferred to G, their virtual velocities would be GK, GI. It is to be proved that $GK = AB$. We have

$$v = AB = AC \cos BAC = MAa \cos BAC = MAa \cos AMS =$$
$$MAa \frac{MS}{MA} = MSa$$

And $m = GK = GH \cos HGK = MG^a \cos HGK = MGa \cos GMS =$
$MGa \dfrac{MS}{MG} = MSa$; whence $m = v$. In like manner it could be shown that $GI = DE$, or $n = v'$. Hence substituting these values in (82),

$$RV = Fv + F'v'. \quad (83.)$$

Now imagine R to be combined in the same manner with one of the remaining forces F'', call R' the resultant of these, V' its virtual velocity. By the same course of reasoning as that just employed,

$$R'V' = RV + F''v'' = Fv + F'v' + F''v''. \quad (84.)$$

In like manner,

$$R''V'' = R'V' + F^n v^n = Fv + F'v' + F''v'' + F^n v^n = \Sigma Fv. \quad (85.)$$

But if there is equilibrium among the forces acting upon the body $R'' = 0$, whence in that case $\Sigma Fv = 0$. (86.) Hence the theorem of virtual velocities is true whether the forces act upon the body at a single point, or have several points of application.

The theorem also holds when the forces are in different planes, but the demonstration of this case is too complicated for the purpose of the present work.

142. Conversely, *if* $\Sigma Fv = 0$, *there will be equilibrium among all the forces.* For from (85) $\Sigma Fv = R''V''$, whence if $\Sigma Fv = 0$, $R'' = 0$. In this case the forces must either balance or form a couple. If they form a couple, the addition of an equal and opposite couple would produce equilibrium. Let P, P', u, u' be the forces and virtual velocities of such an equal and opposite couple. By the proposition $\Sigma Fv + Pu + P'u' = 0$, hence $Pu + P'u' = 0$. As P and P' are parallel forces having different points of application, this equation can hold only when both forces $= 0$.

The theorems of Virtual Velocities are of great practical importance, as they comprehend all the principles of the equilibrium of forces. A valuable application of them will be explained in treating of the mechanical powers (Chapter X.).

CHAPTER IX.

CENTRE OF GRAVITY.

143. Definition. All bodies tend towards the earth by virtue of their weight. Since this tendency is impressed upon every particle, a body is really acted upon by a system of parallel forces having a vertical direction. The resultant of all these must equal their sum, which is evidently the weight of the whole mass. The centre of the system remains at the same point, whatever may be the direction of the forces relatively to the body (§ 120, p. 64), that is, in whatever position it may be placed. This centre is called the *centre of gravity* of the body.

Hence the *centre of gravity* of a body, or system of bodies, may be defined as the point through which the resultant of the weights of the component particles always passes, whatever position the body or system may occupy.

In considering masses as acted upon by gravitation, the whole weight may be supposed to be concentrated at the centre of gravity, so that if this point be supported the body will therefore be balanced about it. Owing to these facts, a knowledge of the location of this point is often of great importance in mechanical problems, and we therefore proceed to a consideration of the methods of determining its position in various cases.

144. Centre of Gravity of Lines.[1] *The centre of gravity of a material straight line is at its middle point*, for the resultant of the weights of the two extreme particles lies at a point midway between them. The same is true for the next two, and thus we may proceed with all the particles.

The centre of gravity of two straight lines inclined at an angle, lies upon the line joining their middle points. Thus let AB, AC, Fig. 59, be two lines united at A. Let G' be the middle point of AB, G'' of AC. The whole weight of AB may be considered as concentrated at G', that of AC at G''. Joining G', G'', the centre of gravity G of the whole system will evidently lie upon the line $G'G''$. To find the position of G we have merely to determine the point of application of the resultant of the weights of AB, AC. Since the moments of these weights about G must be equal,

$$G'G : G''G :: \textit{weight } AC : \textit{weight } AB. \quad (87.)$$

The centre of gravity of a curved line, EBF, Fig. 60, lies somewhere within the segment ABC.

[1] In strict language, we should speak of the centre of gravity of a thin rod instead of the centre of gravity of a line, and of a thin sheet instead of a surface, but as the weights of portions of a homogeneous prismatic or cylindrical rod are proportional to their length, and the weights of sheets of uniform thickness to their surface, we may for the sake of brevity make use of the former terms without inconvenience, provided that their true meaning be kept in mind.

145. Centre of Gravity of Surfaces. *The centre of gravity of any plane surface which is symmetrical about a line lies upon that line.* For let $ABCD$, Fig. 60, be any such surface symmetrical about BD. As there is the same quantity of matter on either side of BD, the centre of gravity of the surface must lie upon it. The line BD is called an axis of symmetry.

If any surface has two or more axes of symmetry, its centre of gravity lies at the intersection of those axes. Let $ABCD$ be such a surface, symmetrical about BD and AC. From what we have already said, it is evident that the centre of gravity lies on both BD and AC. Hence it is at G, their intersection.

The centre of gravity of a parallelogram is therefore at the intersection of its diagonals; that of a circle, ellipse, or any polygon, at its centre of figure. The centre of gravity of a ring is at its geometrical centre, and the same is true of the perimeters of all polygons.

146. Centre of Gravity of a Triangle. *The centre of gravity of a triangle is on a line joining the vertex with the middle of the base, at a distance from the vertex equal to two-thirds the length of that line.* Let ABC, Fig. 61, be a triangle. From A draw AD to the middle point D of BC, also from B draw BE to the middle point E of AC. The lines AD, BE, each divide the triangle into two equal parts; hence the centre of gravity must lie on each line, and is therefore at G, their point of intersection. To determine the position of G, draw ED. ED is parallel to AB because it divides AC, BC, proportionally. Hence the triangles ABC, EDC, are similar, and

$$ED : AB :: DC : BC.$$

Also as ABG, EDG are similar,

$$ED : AB :: GD : GA,$$

whence

$$DC : BC :: GD : GA.$$

But $DC = \frac{1}{2}BC$ by construction, hence $GD = \frac{1}{2}GA = \frac{1}{3}AD$, or $AG = \frac{2}{3}AD$.

147. Centre of Gravity of any Polygon. The centre of gravity of any polygon may be found by dividing it into triangles. Thus let $ABCDE$, Fig. 62, be a polygon. Drawing BE, BD, it is divided into three triangles whose centres of gravity are at g, g', g''. Join g, g'. The centre of gravity, g''', of the portion $ABDE$, must lie on gg', and may be found from the proportion $gg''' : g'g''' :: EBD : ABE$, as the weights of EBD, ABE, may be supposed to be concentrated at g, g'. Now join $g'''g'$. The centre of gravity of the polygon $ABCDE$ must lie on this line at a point G, so situated that

$$g'''G : g''G :: ABE + EBD : DBC. \quad (88.)$$

This process can evidently be applied to polygons having any number of sides, and G may be found either graphically or by trigonometrical calculation.

148. Centre of Gravity of Solids. In the case of homogenous solids symmetrical about any plane, the centre of gravity evidently lies in that plane. Hence the planes of symmetry determine the position of the centre of gravity. If a solid has two planes of symmetry its centre of gravity lies in their line of intersection, if three, at their point of intersection. Hence the centre of gravity of a cube, sphere or any regular polyedron is at its centre of figure.

The centre of gravity of lines, surfaces or solids, with the exception of the most simple ones, are best found by analytical methods requiring the use of the Calculus.

149. Centre of Gravity of Pyramid. As it is frequently necessary in physical problems to know the position of the centre of gravity of a pyramid, we give a geometrical demonstration.

The centre of gravity of a pyramid lies in the line joining the vertex with the centre of gravity of the base, at a distance from the vertex equal to three-fourths the length of that line.

a. Triangular Base. Let $ABCD$, Fig. 63, be a pyramid having a triangular base ABC. Draw AE, BF from the vertices A, B, of the triangles ADC, BDC to the middle of their common base CD. Let g be the centre of gravity of BDC, g' that of ADC. Draw Ag, Bg'. The centre of gravity of the pyramid lies in Ag, for the pyramid may be considered as built up of a series of triangular prisms with bases parallel to BDC, and having a very small altitude; the centre of gravity of each of these lies in Ag, hence the centre of gravity of the whole pyramid lies in that line. It also lies in Bg', since B may be taken as the vertex, and ACD as the base of the pyramid, which may then be considered as composed of a series of prisms with bases parallel to ADC, and having infinitesimal altitudes. Hence as Ag, Bg' are in the same plane ABE, it must be at G, their point of intersection. To determine the position of G join gg'. As gGg', AGB are similar triangles,

$$gg' : AB :: Gg : GA.$$

And as gEg', BEA are similar,

$$gg' : AB :: gE : BE.$$

Combining these proportions,

$$gE : BE :: Gg : GA$$

But $gE = \frac{1}{3} BE$, therefore $Gg = \frac{1}{3} GA = \frac{3}{4} Ag$.

b. Any Base. If the base of the pyramid is not a triangle let it be any polygon whatever, $BCDEF$, Fig. 64. Divide the polygon into triangles BCD, BDE, etc., and pass planes ABD, ABE through the vertex A, and the diagonals BD, BE. The whole pyramid is thus divided into a series of triangular pyramids. Let g be the centre of gravity of the polygonal base, and draw Ag. The centre of gravity of the pyramid is upon this line, since that solid may be supposed to be built up of a series of prisms with bases parallel to $BCDEF$, and of very small altitude, each of whose centres of gravity (and hence the centre of gravity of the whole solid) would be on Ag. It also lies in the plane of the centres of gravity of the triangular pyramids, into which the whole pyramid has been decomposed, that is, in a plane parallel to the base $BCDEF$, and cutting Ag at a distance of three-fourths its length from A. Hence it must be at the intersection of that plane with Ag, or at a point G so situated that $AG = \frac{3}{4} Ag$.

150. Centre of Gravity of Cone. Since a cone may be considered as a pyramid with an infinite number of faces, it follows that the centre of gravity of a cone is situated on the line joining its vertex with the centre of its base, at a distance from the vertex equal to $\frac{3}{4}$ the length of the line.

151. Centre of Gravity of Systems of Bodies. The centre of gravity of connected systems is readily found by the application of the preceding principles.

Let A, B, Fig. 65, be a system composed of two bodies. The mass of A may be considered as concentrated at its centre of gravity g, that of B at g'. The centre of gravity of the system must evidently lie at some point G on gg', so situated that the moments of A and B relatively to it are equal. Hence its position may be determined from the proportion

$$Gg : Gg' :: \textit{weight } B : \textit{weight } A. \quad (89.)$$

In the case of a system containing more than two bodies the position of the centre of gravity may be determined by a process similar to that used in the case of an irregular polygon (§ 146, p. 76). The application of these principles to extended systems of bodies is of great importance in astronomy.

152. Centre of Gravity of Two Bodies connected by a Homogeneous Rod. To find the centre of gravity of two bodies connected by a rod we first find the centre of gravity of the bodies themselves, and then find the centre of gravity of the rod. By considering the weight of bodies as concentrated at their centres of gravity, and that of the rod as concentrated at its middle point, we readily determine the centre of gravity of the whole system. Thus let A, B, Fig. 65, be the bodies, and gg' the homogenous rod connecting them. The centre of gravity of A and B is at G, that of gg' at C. The centre of gravity of the whole system is at x, so situated that

$$Cx : Gx :: \textit{weight of } A + B : \textit{weight of } gg'. \quad (90.)$$

153. Distance of Centre of Gravity of two or more Bodies from a Plane. *The distance of the centre of gravity of any number of bodies from a plane is equal to the quotient arising from dividing the product of the weight of each body into the distance of its centre of gravity from that plane, by the sum of the weights of the bodies.*

This follows directly from equations (70), (71), (72), p. 73.

For let ZOY, Fig. 56, be the plane in question. The weights of the bodies may be represented by F, F', F'', F^n, and considered as parallel forces applied at points (x, y, z), (x', y', z'), etc. The distances of the centres of gravity of the separate bodies from ZOY are x, x', x'', x^n. The centre of gravity of the system of bodies will therefore be the centre of these parallel forces, which lies at a distance x_r from ZOY, so situated that $x_r = \frac{\Sigma F x}{\Sigma F}$.

154. General Analytical Methods. The equations of § 136 (70, 71, 72) furnish a general method of determining the position of the centre of gravity of a system of bodies by reference to trilinear coördinates. If we substitute for F, F', etc., in those equations the quantities M, M', etc., the weights of the bodies, and denote by (x, y, z), (x', y', z'), etc., the coördinates of their centres of gravity, the coördinates x_r, y_r, z_r will give the position of the common centre of gravity of the system, this point being the centre of a combination of parallel forces M, M', M'', M^n applied at points (x, y, x), (x', y', z'), etc. Hence we obtain the following equations, in which x_r, y_r, z_r are the coördinates of the centre of gravity of the system of bodies M, M', M'', M^n,

$$x_r = \frac{\Sigma Mx}{\Sigma M} \quad (91.) \qquad y_r = \frac{\Sigma My}{\Sigma M} \quad (92.) \qquad z_r = \frac{\Sigma Mz}{\Sigma M} \quad (93.)$$

155. Application of the Calculus. The processes of integral calculus enable us to apply the preceding method to the determination of the position of the centre of gravity in the case of a body of any regular geometrical form, as all bodies may be considered as systems of particles. Equations (91), (92), (93) are evidently true, whatever the magnitude of the masses composing them; hence, if we imagine a body to be divided into a series of infinitesimal weight-elements, they still hold. Suppose this division to take place in the case of any body, and let dM be the element of weight, x, y, z, the coördinates of its centre of gravity. The moments of any weight-element relatively to the axes Z, X, Y respectively are xdM, ydM, zdM, and the sum of the moments represented in (91), (92), (93), by ΣMx, ΣMy, ΣMz, must be replaced by $\int xdM$, $\int ydM$, $\int zdM$, as dM is an infinitesimal. Also ΣM must be replaced by $\int dM$. Hence for any body whatever the coördinates of the centre of gravity are

$$x_r = \frac{\int xdM}{\int dM}, \quad (94.) \qquad y_r = \frac{\int ydM}{\int dM}, \quad (95.) \qquad z_r = \frac{\int zdM}{\int dM}. \quad (96.)$$

If the body under consideration is homogeneous dM is equal to its volume, which is a volume element dV, multiplied by the weight of a unit of mass. As this latter factor is a constant, for homogeneous bodies

$$x_r = \frac{\int xdV}{\int dV}, \quad (97.) \qquad y_r = \frac{\int ydV}{\int dV}, \quad (98.) \qquad z_r = \frac{\int zdV}{\int dV}. \quad (99.)$$

The manner in which the body is supposed to be divided to form weight-elements is different in different cases, as certain modes of imaginary division oftentimes simplify the problem very greatly. The integrations must of course be taken between *limits*, as indicated by the conditions of the problem to be solved. For the practical application of the preceding principles the student is referred to the standard treatises on analytical mechanics.

156. Theorems of Pappus. An interesting application of the principles which we have demonstrated in the preceding pages is found in the *centrobaryc method* of determining the volumes of solids of revolution, and the contents of surfaces of revolution. The principles of this method were first stated by Pappus of Alexandria (380 B. C.), and are hence called after him. They are also known as the *Principles of Guldinus*, from a mathematician who republished them about 1640 in a work on centres of gravity of lines, surfaces and solids. The first real demonstration of them, however, appeared in 1647 in a work by Cavalieri.[1] They are as follows:—

I. *The volume of the solid generated by the revolution of a surface about an axis in the same plane with it, is equal to the area of the surface multiplied by the circumference described by its centre of gravity.*

II. *The area of the surface generated by the rotation of a line about an axis lying in the same plane with it, is equal to the product of the line into the circumference described by its centre of gravity.*

Let ABC, Fig. 66, be a surface generating the solid, of which $ABC-DEF$ is a part, by its revolution about the axis X. Any element of ABC, as P, generates a solid by its revolution, as indicated by the dotted lines, and the sum of the solids thus formed by the elements, P, P', etc., of which ABC is composed, is evidently equal to the whole solid of revolution. Supposing these elements to be very small, and denoting by r, r', etc., their distances PR, $P'S$ from the axis X, the prismatic solid described by $P = 2\pi r \times P$; that described by $P' = 2\pi r' \times P'$, and so on, for all the elements. Hence the volume of the whole solid $=$ $2\pi r \times P + 2\pi r' \times P' + 2\pi r'' \times P'' \ldots + 2\pi r^n \times P^n =$ $2\pi(Pr + P'r' + P''r'' \ldots + P^n r^n)$. Now if the distance of the centre of gravity of ABC from x be denoted by r_0, we have from the demonstration of § 152, p. 78, $Pr + P'r' + P''r'' \ldots + P^n r^n$ $= r_0(P + P' + P'' + \ldots P^n)$. If each of these equals be multiplied by 2π, we have $2\pi(Pr + P'r' + P''r'' \ldots + P^n r^n) =$ $2\pi r_0(P + P' + P'' \ldots + P^n)$. But $P + P' + P'' \ldots + P^n =$ volume of ABC, $2\pi r_0$ is the circumference described by its centre of gravity, and the first member of the equation represents the sum of the solids generated by the elements of ABC, that is, the whole solid of revolution.

Theorem II. follows directly from Theorem I., since a line may be considered as a surface of infinitesimal width, in which case the solid of revolution generated becomes a surface of revolution.

If the surface ABC revolves only through an angle a instead of making a complete revolution, the volume of the solid generated

$$= 2\pi r_0 \times \textit{area of surface} \times \frac{a}{360}.$$

[1] See *The Theorem of Pappus*, by J. B. Henck; *Mathematical Monthly*, Vol. I., p. 200.

As an illustration of the preceding let us take the case of the *torus* formed by the revolution of a circle about X. Let the area of the circle be πR^2, and r_0 the distance of its centre from X. Then the volume of the torus generated $= 2\pi r_0 \times \pi R^2$.

Again let it be required to find the area of the circle generated by the revolution of its radius R about X, by the centrobaryc method. The length of the revolving line is R, the distance of its centre of gravity from X, $r_0 = \frac{R}{2}$. Hence the surface generated $= 2\pi \frac{R}{2} \times R = \pi R^2$.

Equilibrium of Bodies as affected by Position of Centre of Gravity.

157. Case of Suspended Bodies. Since a body is upheld when its centre of gravity is supported, the nature of the equilibrium of a body is determined by the position of that point relatively to the points of support. We consider first the case of suspended bodies.

That any suspended body may be in equilibrium, its point of support and centre of gravity must lie in the same vertical, because the resultant weight is then applied directly at the point of suspension, and is balanced by its reaction. Three cases may arise: 1, when the centre of gravity is below the point of suspension; 2, when it is above the point of suspension; 3, when the centre of gravity and point of suspension coincide.

158. Stable Equilibrium. If the centre of gravity is below the point of suspension, the body will return to its former position of equilibrium on being removed from it. The equilibrium in this case is therefore *stable*. Thus let MN, Fig. 67, be a body suspended at S. Let G' be the position of the centre of gravity when removed from its position of equilibrium, so that S and G' are not in the same vertical. The total weight of the body acts directly downward through G', and may be represented by $G'A$. This force may be resolved into two others represented by $G'B$, $G'C$, the former of which simply produces pressure on the pivot at S, by the reaction of which it is balanced, while the latter produces a rotary motion about S as an axis. Hence the body can not come to rest until $G'C$ becomes equal to zero, which occurs when G' takes the position G in the same vertical with S.

159. Unstable Equilibrium. *Unstable equilibrium* occurs when the centre of gravity is above the point of suspension. In this case, if the body be moved so that the two are no longer in the same vertical, it will not return to its former position, but will revolve about the point of suspension till the centre of gravity reaches the lowest possible position. This will be seen by an inspection of Fig. 68, in which $G'A$ represents the weight of the body, $G'B$, $G'C$, its components as before.

160. Neutral or Indifferent Equilibrium. A body suspended at its centre of gravity will remain at rest indifferently in any position, the total weight being always applied at the point of suspension, however the body may be moved. This is *neutral equilibrium.*

The various conditions of equilibrium may be illustrated experimentally by suspending a wheel successively on axes passing above, below and through its centre.

161. Experimental Method of finding Position of Centre of Gravity. From what precedes it appears that the equilibrium of any suspended body is stable only when the centre of gravity is at the lowest possible point, so that if any body be suspended freely its centre of gravity will assume this position. We are thus furnished with a method of finding the centre of gravity by experiment. For example, suppose it is required to find this point in the case of a thin sheet of metal *MN*, Fig. 69. If it be suspended from a point *S* the centre of gravity will assume the lowest possible position. Hence if a plumb-line *SB* also be hung from *S* the centre of gravity of the surface of *MN* must be somewhere on the line *ST*. Marking this line on *MN*, let the body be suspended from some other point, as *S*. The centre of gravity of the surface will now be on *SB*, and as it is also on *ST*, it will be at *G*, the intersection of these two lines. The centre of gravity of the plate must therefore be on a perpendicular to the surface passing through *G*, at a depth equal to half the thickness of the plate.

162. Measure of Stability of Suspended Bodies. The stability of any suspended body is measured by the force required to hold it in an inclined position, which is equal to the force with which it tends to reassume its position of stable equilibrium. The *stability is proportional to the weight of the body multiplied by the distance of the centre of gravity below the axis of suspension.* Let *MN*, Fig. 70, be the body under consideration. *G* its centre of gravity, and S the axis of suspension. The total force tending to move it is the moment of its weight, or $W \times SP$. If now we suppose the centre of gravity to be lowered to G', and the weight to be changed to W', the new moment $= W' \times SP'$. Hence calling M and M' the stabilities in the two cases, we have $M : M' : W \times SP : W' \times SP'$, or as $SP : SP' :: SG ; SG'$,. $M : M' :: W \times SG : W' \times SG'$. (100.)

163. Equilibrium of Bodies resting on a Horizontal Plane. A body resting upon a horizontal plane is in equilibrium whenever the vertical passing through the centre of gravity falls within the base, as in that case the weight is directly opposed by the reaction of the supporting surface. Thus if *G* be the centre of gravity of *AB*, Fig. 71, the body will stand, as *GM* falls within the base; but if it be raised to *G'* by the addition of the piece *AC*, the body will be overturned as *G'M'* falls without the base. The vertical through the centre of gravity is often called the *line of direction.* The truth of this proposition may be verified by suspending a plumb-line from *G* and *G'*. If the base of the body be curved, the line of direction must evidently pass through the point in which this base touches the plane on which it rests; otherwise the body will roll until that position is assumed.

The base of a body resting upon a point is evidently that point; if the body is supported by two points, the line joining them may be considered as the base; if on three or more points, the base is the polygon included between the lines joining the points of support.

164. Illustrations and Applications. So long as the vertical through G lies within the base the body is secure. Thus the leaning towers of Pisa and Bologna have stood for centuries, because although they deviate from the perpendicular by a considerable amount, the line of direction still falls so far within their bases that no ordinary shock would be sufficient to overthrow them. The height of the tower at Pisa (which is the *campanile* or bell-tower of the cathedral) is in round numbers 87 m., and the diameter of its base is 17 m. It is inclined 4.7 m. from the perpendicular, but the vertical through the centre of gravity falls 3 m. within the base. The leaning tower of Asinelli at Bologna is 100 m. in height, with an inclination of 1 m. The Garisenda tower in the same city is 63 m. high. with an inclination of 3 m.[1]

The principle explained in the first part of § 163 furnishes another experimental method of obtaining the centre of gravity of a body. This may be done by balancing it on its edge, in which case the centre of gravity must lie in the vertical plane passing through that edge. Balancing a body in three different positions determines its centre of gravity, which must be the point of intersection of three vertical planes passed through the edge in three successive positions of the supported body.

The equilibrium of a body bounded by plane surfaces is stable when the body rests upon a side, and unstable when it rests upon an angle. If a body be rolled it assumes positions of stable and unstable equilibrium alternately. The equilibrium of a body resting on a horizontal surface is evidently *indifferent* so far as motion in a horizontal plane is concerned.

165. Body with curved Surface. — Metacenter. The nature of the equilibrium of a body having a curved surface is determined by the relative position of the centre of gravity and a point known as the *metacenter*. Let ACD, Fig. 72, be a solid bounded by a curved surface, and resting on a plane RQ. That it may be in equilibrium the axis AB must be vertical, as only in that position will the line of direction fall within the base. Suppose now that the body be moved slightly to some new position in which AB is not vertical. It will then be acted upon by two forces, the resultant weight acting downward through G, and the reaction of the surface RQ acting upward along the line PM. The point M, in which AB and PM intersect, is called the *metacenter* of the body.

The forces through G and M form a couple, as indicated by the arrows, the tendency of which is to make ACD return to its orig-

[1] Considerable discrepancy exists among these figures, as given by different authorities. The data above are reduced from *Appleton's American Encyclopedia*.

inal position, so long as the centre of gravity is below M. If it assumes the position G', above M, the tendency of the couple is to overturn the body, and the equilibrium of the body in its original position is unstable. If M and G coincide, the body remains at rest in any position.

Hence the equilibrium is *stable*, *unstable* or *indifferent*, according as the centre of gravity is *below*, *above* or *coincident with* the metacenter, or, in general, *the equilibrium of a body with a curved surface is the same as if it were suspended at its metacenter.*

166. Equilibrium of a Body resting on Inclined Plane. Under these circumstances there is equilibrium when the line of direction falls within the base, precisely as in the case of a body resting on a horizontal surface.

167. General Criterion of Equilibrium. The different conditions of equilibrium demonstrated in the preceding paragraphs may all be comprehended in a single proposition, as follows: — *In whatever manner a body is supported, the equilibrium is stable when on moving it the centre of gravity ascends, unstable when the centre of gravity descends, and indifferent if it neither ascends nor descends.*

This will be seen to be true of suspended bodies by a simple inspection of Figs. 67, 68. In the case of a body resting on a base, as AB, Fig. 71, G will evidently rise if AB is slightly rotated about either edge of its base. If AB rested on an angle, as B, the equilibrium would be unstable, and G would then occupy its highest possible position, so that on disturbing the body G would move in a descending curve. A consideration of Fig. 72 will show that when a body rests on a curved base, if the centre of gravity be below M, as at G, it moves in an ascending curve on disturbing the body, while if it is at G' above M, it moves in a descending curve, and if at M the motion is in a horizontal line.

If the surface of the body possesses a different curvature in different planes, the body may be in stable, unstable, or indifferent equilibrium, according to the portion of its surface on which it rests. Thus an ellipsoid of revolution is in unstable equilibrium with regard to motion in any plane if placed on either extremity of its longer axis, while if placed on its shorter axis its equilibrium is stable with regard to motion in the vertical plane passing through its major axis, and indifferent with regard to any plane at right angles to this.

168. Experimental Illustrations. The preceding principles explain the deportment of the toys so frequently seen, which represent a grotesque human figure mounted on a curved base. The figure persistently resumes its upright position when laid upon its side. This is explained by the form of the curved surface constituting the base, which is so constructed that the centre of gravity shall be in its lowest possible position when the figure is upright. A method sometimes used to cause an egg to stand on one end is also capable of a similar explanation. The yolk of an egg is contained in a sac nearly concentric with the white, and is somewhat denser than the latter. By a vigorous shaking the yolk-sac may be broken, and the heavy liquid then falls to the lower end of the egg when this is set upright, thus bringing the centre of gravity to a point so low that any motion of the egg tends to raise it. Another experiment often shown is that of causing a wooden cylinder to roll up an inclined plane. This is done by weighting a portion of the cylinder with lead, so that the centre of gravity does not coincide with the centre of magnitude. The cylinder can easily be so placed that the rotation produced by the tendency of the centre of gravity to assume the lowest possible position at the same time causes the body to ascend the plane. The slope of the plane must of course be so small that the rise of the centre of gravity caused by the ascent of the plane shall be less than the fall due to the rotation of the cylinder. A similar explanation applies to the case of two cones united at their bases, which can be made to roll up two inclined planes set at a suitable angle. In this case, as in the last, the centre of gravity really falls, since the rise caused by the rolling of the cones up the plane is less than the fall produced by the simultaneous approach of the points of support to their vertices.

169. Practical Illustrations of Equilibrium. All the attitudes and movements of animals are regulated with regard to the position of the centre of gravity of their bodies. In man, when erect the centre of gravity is about in the median line of the body, but its place changes with every change of position. The whole art in feats of posturing consists in keeping the line of direction within the base formed by the feet. If a limb be extended in any direction the centre of gravity moves in the same direction, but if some other portion of the body be thrown toward the opposite side the equilibrium is preserved. When a porter carries a load upon his back he must bend forward, while a nurse carrying a child in her arms leans backward, otherwise the line of direction falls in one case behind the heels, in the other in front of the toes. For a like reason when climbing a steep hill we bend forward, and on descending it we lean backward. The art of walking on stilts depends on the ability to keep the centre of gravity directly over the narrow base formed by them. The same is true of the art of walking or dancing on the tight rope. In this case the performer generally carries a heavy pole, by moving which in any direction he can more readily bring the centre of gravity to the desired position, thus making his feats far easier and more graceful than if unassisted by this aid. When a quadruped progresses it never raises both feet on one side simultaneously, for in that case the centre of gravity would be unsupported, but the fore foot on one side and the hind foot on the other are lifted together in trotting, the fore foot a little in advance in walking, or else, as in galloping, the two fore legs are lifted, and the body projected forward by the spring of the hind legs. The principles of equilibrium explain the different feats of balancing long poles, swords, etc., in an upright position upon the hand. Feats of this kind are made much easier if a horizontal rotation is given to the body which is balanced, because the centre of gravity describes a small circle about a fixed point, so that even if it is not directly over the base the direction in which it tends to fall is constantly changing, causing a preservation of equilibrium. It is thus that a top remains vertical when spinning rapidly, but falls when at rest. The same explanation applies to the common experiment of balancing a plate upon the point of a sword. This is impossible if the plate is at rest, but if it be made to whirl rapidly no difficulty is experienced.

170. Measure of the Stability of Bodies resting on a Plane.—Statical Stability. We may consider the stability of a body under two aspects: 1st, in relation to the force necessary to move it from its position of equilibrium; 2d, in relation to the force required to completely overturn it. The forces exerted in these two cases determine the *statical* and *dynamical stability of a body*.

Statical Stability. This is measured by the force necessary to begin rotation about an angle of the body. Let MN, Fig. 73, be a mass resting on a horizontal plane, and suppose a force F to be applied along the line CF. If any obstacle be placed at N, so that MN cannot slide along the plane, F will tend to rotate MN about N. The moment of F relatively to an axis through N is $F \times DN$. The moment of the weight W of the body tending to prevent motion is $W \times PN$. When motion begins, $F \times DN = W \times PN$. Hence $W \times PN$ measures the statical stability of MN, and this is jointly proportional to the weight of the body and to the distance of the foot, the vertical through its centre of gravity from the axis about which rotation tends to take place. Hence the greater the weight and the broader the base the greater the statical stability of a body, which is, however, independent of the height of the centre of gravity. If the distance PN is different in different directions, the stability of MN varies according to the direction of the force F.

The product $W \times PN$ is called the *moment of stability* of the body MN.

171. Dynamical Stability. The dynamical stability of a body is measured by the moving force expended in overturning it. To accomplish this in the case of any mass MN, Fig. 74, it must be revolved about one of its angles N, until the centre of gravity, which describes the curve GH, reaches its highest point H, after which the action of gravity will complete the overturning. Hence the whole force is expended in raising G through the vertical height HL. The power required to do this must be proportional to the weight W of the body, and also to the height HL, through which this weight is raised; hence it is measured by their product $W \times HL$. If the centre of gravity occupies a more elevated position G', the distance $H'L'$ will be less than HL, hence less power will in that case be required to overturn the body. Therefore the dynamical stability of a body increases with its weight, and diminishes with the elevation of its centre of gravity.

172. Practical Illustrations of Stability. Many illustrations of the two kinds of stability may be cited. A pyramid owes its great firmness to the breadth of its base and the low position of the centre of gravity. Walls are often strengthened with small expenditure of material by building offsets or by *battering* the side toward which the forces tending to overturn them are directed. A high carriage is upset much more readily than a low one, because the centre of gravity is higher, so that a smaller horizontal force overcomes its dynamical stability, and also if the road is sloping a less inclination is required to throw the line of direction outside of the base. For this reason wagons intended for carrying heavy loads are generally made so as to have the greater part of the load lower than the axles of the wheels. Quadrupeds have a much greater stability than man, because of the broader base formed by their feet, hence their young acquire the art of walking far more rapidly than a child, which must first learn the art of balancing itself on its legs. The human figure is most stable when the feet are so placed as to make the base as large as possible. If the heels are in contact this occurs when the angle of the feet is 90°. If the heels are separated by an amount equal to the length of the foot, their angle should be 60°.

CHAPTER X.

Mechanical Powers.

173. Definition. The forces subject to our use can be applied only by means of machinery of greater or less complexity. Machines are either simple or compound. The elementary machines to which all systems may be reduced are called the *mechanical powers*. They are seven in number, viz.:—the *lever*, *wheel and axle*, *cord*, *pulley*, *inclined plane*, *wedge*, and *screw*. These may again, in principle, be reduced to three, the *lever*, *cord* and *inclined plane*, the first of these comprehending the *lever* and *wheel and axle*, the second the *cord* and *pulley*, the third the *inclined plane*, the *wedge* and the *screw*.

The force applied to a machine is called the *power*; the opposition which it is to overcome is the *weight* or *resistance*.

I. *Lever.*

174. Definition. A *lever* is a rod either crooked or straight, moving about a fixed point called the *fulcrum*. The theoretical lever, which we proceed to consider, is supposed to be without weight, and absolutely inflexible.

For convenience of description, levers are divided into three classes, according to the relative position of the power, weight and fulcrum. When the fulcrum is between the power and the weight, the lever is said to be of the *first order*, Fig. 75; if the weight is between the power and fulcrum, the lever is of the *second order*, Fig. 76; and if the power is between the weight and fulcrum the lever is of the *third order*, Fig. 77. In the two last classes the power and weight act in opposite directions; in the first they act in the same direction.

175. Conditions of Equilibrium. There is equilibrium with the lever when the moments of the power and of the weight relatively to the fulcrum are equal. Thus in the levers represented in Figs. 75, 76, 77, 78, equilibrium occurs when $P \times AF = W \times BF$, or when $P : W :: BF : AF$. (101.)

From (101) it follows that $W = P\frac{AF}{BF}$ (102.), which shows that there is, in general, a mechanical advantage gained in employing levers of the first and second orders, while in those of the third order the power acts at a disadvantage.

When the lever is straight, and the forces act at right angles to it, as in Figs. 75, 76, 77, the arms of the forces evidently coincide with the portions of the lever lying between them and the fulcrum.

If any number of forces act upon a lever there will be equilibrium when the algebraic sum of their moments relatively to the fulcrum is zero.[1]

176. Pressure upon the Fulcrum. P and W being balanced about the fulcrum, this must sustain a certain pressure due to their coincident action. From § 108 (p. 59) it appears that when P and W are in equilibrium their resultant must pass through F, and the magnitude and direction of this resultant, which is the pressure on F, can be determined, as shown in § 106 (p. 59). If the forces act in parallel directions, the pressure on the fulcrum *in that direction* equals their algebraic sum. The pressure upon the fulcrum in any other direction can be found by resolving the resultant of P and W into two components, one in the required direction, the other at right-angles to it. In case several forces act upon a lever, the pressure on the fulcrum equals the resultant of all of them.

177. Familiar Examples. As familiar examples of the different kinds of levers, we mention the following: *1st Order.* A crowbar used to raise a weight, a hammer applied to draw a nail, and the handle of a common pump. Scissors, pincers, etc., are composed of two levers of the first order, the pivot being the fulcrum, and the substance to be cut the resistance. *2d Order.* An oar is a lever of the second kind, the water serving for the fulcrum, the boat being the weight, and the muscular force of the arm the power. The common vertical hay-cutter, and the machine used by apothecaries for compressing corks, are additional examples. So also is the common wheel-barrow, the wheel being the fulcrum, and the load the weight. *3d Order.* In levers of this class the power is always greater than the weight, since its arm is less than the weight-arm. The treadle of a turning-lathe, or of a grindstone, is a good example. Tongs are also levers of the third order. We have another illustration in the use of a fishing-rod, the fish being the weight, the hand raising the rod the power, and the other hand the fulcrum. The mechanical disadvantage of this kind of lever is well seen in the case of a man raising a long ladder against the

[1] In the present chapter the forces are all supposed to lie in the same plane perpendicular to the axis of the fulcrum.

side of a building. The foot of the ladder is the fulcrum, and the weight may be supposed to be concentrated at its centre of gravity; as the power must be applied near the base, it must exceed the weight of the ladder until the inclination of the latter becomes very slight.

178. Relative Velocity of Power and Weight. A consideration of Figs. 75, 76, 77, 78, will show that if motion ensues about the fulcrum,

$$\textit{Velocity of } P : \textit{Velocity of } W :: AF : BF :: W : P. \quad (103.)$$

Hence the power moves through a proportionally greater distance than the weight in the first two classes of levers, while in the third class a small motion of the power causes a proportionally large motion of the weight.

The latter principle is beautifully applied in the human system. Thus it is necessary to be able to move the hand over a considerable distance with the elbow-joint as a centre of motion, and this must be done by a comparatively slight contraction of the flexor muscle of the fore-arm. One of the extremities of this muscle is attached to the upper part of the humerus, the other to one of the bones of the fore-arm, at a point very near the elbow-joint. Hence when the muscle contracts, the extremity of the arm is carried over a distance greater than the amount of the muscular contraction, in proportion as its length is greater than the distance from the joint to the point of attachment of the muscle. There are also several examples of levers of the first and second classes in the human skeleton.

179. Compound Lever. The compound lever is a system of simple levers, so arranged as to act upon one another successively, as shown in Fig. 79. The power, P, being applied at the extremity of the arm L of the first lever, the arm l acts upon L' of the second lever, l' in its time acts upon L'', and so on. The weight is applied at the extremity of l''. Supposing the forces to act at right angles to L, l'', the efficiency of the compound lever may be found as follows: The pressure acting on $L = P$, whence by (102.), p. 88, it follows that the pressure acting on L' or $W = P\frac{L}{l}$. In like manner, the pressure on $L'' = W' = W\frac{L'}{l'} = P \cdot \frac{L}{l} \cdot \frac{L'}{l'}$. In the same way we find the pressure exerted by l'' to sustain the weight W'', is $W'' = W'\frac{L''}{l''} = P\frac{L}{l} \cdot \frac{L'}{l'} \cdot \frac{L''}{l''}$ (104.), whence

$$P \times L \times L' \times L'' = W'' \times l \times l' \times l''. \quad (105.)$$

If the forces do not all act at right-angles to the levers, L, L', L'', l, l', l'' must be considered as the *arms* of the forces P, W, W', W'', that is, the perpendiculars from F, F', F'', dropped upon their lines of action. It is evident that the above demonstration applies equally to all three orders of levers. Hence *there is equilibrium with a compound lever when the power multiplied by the continued product of the power-arms is equal to the weight multiplied by the continued product of the weight-arms.*

180. Effect of Weight of Lever. Hitherto we have neglected the weight of the lever altogether, which it is not generally admissible to do in practice. It remains, therefore, to explain the method of taking this into account. The weight of the lever may always be regarded as an additional force applied at its centre of gravity and acting vertically downward. For prismatic and cylindrical levers, the formulæ may be deduced in the following manner. Call w the weight of the lever in all cases. As this acts through g, the middle point of the lever, the conditions of equilibrium are that the algebraic sum of the moments of P, W, and w, relatively to F, shall be zero. Calling the power-arm L, the weight-arm l, and denoting right-handed moments by $+$, left-handed by $-$, the algebraic expression of these conditions for levers of the first order is,

$$-PL - w\,\tfrac{1}{2}(L-l)^1 + Wl = 0, \text{ whence } W = \frac{PL + w\,\frac{1}{2}(L-l)}{l} \quad (106.)$$

For levers of the 2d order,

$$PL - (Wl + w\,\tfrac{1}{2}L) = 0, \text{ whence } W = \frac{PL - \frac{1}{2}wL}{l} \quad (107.),$$

and for levers of the 3d order,

$$PL - (Wl + w\,\tfrac{1}{2}l) = 0, \text{ whence } W = \frac{PL - \frac{1}{2}wl}{l}. \quad (108.)$$

181. Body resting on Props. An important application of the principle of the lever is the determination of the pressure sustained by each of two or more props supporting a loaded beam. Let AB, Fig. 80, be a beam resting on two props, A, B, and sustaining a weight W. Suppose B to be a fulcrum, about which W tends to turn the beam AB. The pressure exerted on A will be equal to the power which must be applied at A to hold W in equilibrium. Call this power P. Then from (101.),

$$P : W :: BC : AB, \text{ whence } \textit{Pressure on } A = P = W\frac{BC}{AB}. \quad (109.)$$

In like manner, supposing A to be the fulcrum, the pressure on B, (P') is found from the equation,

$$P' : W :: AC : AB, \text{ whence } \textit{Pressure on } B = P' = W\frac{AC}{AB}. \quad (110.)$$

The total pressure found by adding these two together is

$$P + P' = W\left(\frac{BC}{AB} + \frac{AC}{AB}\right) = W,$$

as might be inferred from the laws of parallel forces.

In this demonstration we have not taken into account the weight of the beam AB. When this is necessary equation (107) may be used by considering a power (P) applied at A and B successively, to balance $W + w$, and solving relatively to P.

182. Balance with Equal Arms.

Prominent among the applications of the lever are the various forms of instruments used for weighing, as the common balance, steelyard, platform-scales, etc. We proceed to explain the most important of these.

The common balance consists essentially of a lever of the first order, AB, Fig. 81, supported at its middle point F by means of

[1] For $W \times gF$ = moment of w relatively to F. But $gB = \frac{1}{2}AB = \frac{1}{2}(AF + FB) = \frac{1}{2}(L + l)$. Hence $Bg - BF = gF = \frac{1}{2}(L + l) - l = \frac{1}{2}(L - l)$.

a knife-edge, and having two scale-pans *C*, *D*, suspended from its extremities. In one of these pans is placed the substance to be weighed, in the other the standard weight necessary to balance it. The lever *AB* is called the *beam* of the balance, *F* the fulcrum, and *A*, *B*, the *points of suspension* of the pans. An index *I* is generally attached to show when the beam is horizontal.

183. Characteristics of a good Balance. A good balance must possess the following characteristics: 1, *truth*, that is, it must be so adjusted that the beam shall be horizontal when the weights in the two pans are equal; 2, *stability*, the beam should tend to return to its horizontal position when deflected from it; 3, *sensibility*, the beam should be deviated from its horizontal position by a very slight excess of weight in either pan, and the greater the deviation with a given weight, the greater the sensibility.

To ensure *truth* (1), the two arms of the balance must be of exactly the same length and weight, otherwise the weights in the pans would be unequal when the beam is horizontal. To find whether this condition is fulfilled, a body is placed in one pan, and in the other the weight necessary to balance it. The contents of the pans are then transposed, and if the arms are equal the equilibrium is still maintained. (2.) The scale-pans should evidently be of the same weight, so that the beam may be horizontal when they are empty. (3.) All the parts of the balance should be symmetrical about two planes passing through the centre of gravity of the beam; one of these planes containing the longer axis of the beam, the other being at right angles to it.

To ensure stability, the fulcrum *F*, Fig. 82, and the centre of gravity of the beam *G*, should not coincide, but should both lie in the same perpendicular *FG*, to the line *AB* joining the points of suspension, the centre of gravity being *below* the fulcrum. (1.) *G* should not *coincide* with *F*, for if this were the case the equilibrium of the beam would be indifferent, and hence when inclined it would remain in that position, with no tendency to return to horizontality. (2.) *G* and *F* should be in the same perpendicular to *AB*, for otherwise the centre of gravity would not occupy its lowest position when the beam is horizontal. (3.) *G* should be *below* *F*, for if it were above, the equilibrium would be unstable, and the beam would overturn on being deflected from horizontality.

As the sensibility of a balance depends upon the weight necessary to deflect the beam through a given angle, it is evidently increased by everything which diminishes the friction at the fulcrum and points of suspension. Hence for fine balances it is customary to make the fulcrum *F* of hard polished steel, having a sharp knife-edge, and supported upon a plane surface of agate. The points of suspension are also made of wedges of steel. The three knife-edges should be placed so as to be horizontal, and perpendicular to the plane of the beam. They should also lie in the same straight line AB, for the weights of the bodies placed in the scale-pans may be considered as applied at A and B, and if the centre of these equal and parallel forces is at F, they will produce no tendency to motion about that point, whatever may be the position of the beam, so that the sensibility will be independent of the weights in the pans. If, however, F is above the line AB, the centre of the forces will be below F, and hence will tend to keep the beam horizontal, thus diminishing the sensibility of the balance in proportion to their magnitude. If F were below AB, the equilibrium would be unstable when any considerable weight was contained in the pans.

The other conditions of sensibility may be deduced as follows: Suppose weights P, Q, to be placed in the scale-pans, P being greater than Q. The beam is deflected in the direction of the greater weight, assuming an inclined position, as shown in the figure. (Fig. 82.) There is now equilibrium between the moment of the deflecting force $P - Q$ (the excess of weight in the left hand pan) applied at A, and the weight of the beam, w, applied at G, and tending to restore the beam to horizontality. Hence when the beam comes to rest in an inclined position, the moments of these forces relatively to F must be equal; that is,

$$(P - Q) \times CF = w \times mF.$$

Denote the angle $CFA = mGF$ by a. Then $CF = AF \cos a$, $mF = GF \sin a$. Substituting these values in the above equation,

$$(P - Q) \times AF \cos a = w \times GF \sin a, \text{ whence}$$

$$\text{tang } a = \frac{(P - Q) \times AF}{w \times GF}. \quad (110.)$$

Now the angle a, through which the beam is inclined by a given excess of weight $P - Q$, measures the sensibility, which therefore increases as tang a increases; that is, (1.) as the arm AF is greater, that is, as the beam is longer; (2.) as the weight of the beam w is less; (3.) as the distance of the centre of gravity of the beam below the fulcrum (FG) is less.

Practically these conditions are true only within certain limits. If the beam is too light, or if the arms are very long, they bend under the action of the weights in the pans, thus bringing the centre of these forces below F, and so diminishing the sensibility. Also if G is too near F, a light weight deflects the beam by an inconveniently large amount.

184. Measure of the Sensibility of a Balance. The sensibility of balances may be compared by observing the angle through which their beams are inclined by a weight placed in either pan. Thus if a weight of 1 mgr. incline one balance through half a degree, and another through 1 degree, the second is twice as sensitive as the first. An easier method is by determining the smallest additional weight that will turn the balance perceptibly when this is loaded by a given amount. For example, if a balance carries a load of 1 kgr. in each pan, and the smallest weight capable of deflecting its beam is 1 mgr., the balance is said to be sensible to $\frac{1}{1000000}$ of that load; that is, with such a balance the weight of 1 kgr. can be determined within 1 mgr. Theoretically, the sensibility of a balance constructed as described, is independent of the load, but practically it diminishes as the load increases, because (1.) a slight bending of the beam occurs; (2.) the friction between the knife-edges and their supports is increased by the additional pressure; and (3.) it is practically impossible to get the three knife-edges exactly in line. For these reasons the weight with which the balance is loaded must always be stated when the sensibility is estimated by this method.

Still another way of determining sensibility is by observing the number of oscillations made by the beam in a given time. Whenever the beam is removed from its position of equilibrium, it performs a series of oscillations about that position before coming to rest. The slower these oscillations the greater the sensibility of the balance. The reason of this will be seen from Fig. 82. Suppose the pans of the balance to be empty, and let the beam be moved from its horizontal position until it occupies the inclined position *AB*. The moment of the weight of the beam, which is $w \times GF \sin \alpha$, will now be exerted to cause it to return to horizontality, and it will vibrate about the line *CD* in gradually diminishing arcs, finally coming to rest with its axis coincident with that line. Now it is evident that the less the moment of the impelling force, that is, the less the value of w and GF, and hence greater the sensibility, the slower the oscillations.

A good chemical balance should weigh to $\frac{1}{10}$ mgr. when each pan is loaded with 50 grammes. The best balances detect a difference of load of 1 mgr. with a weight of 2 kgrs. in each pan, a sensibility of $\frac{1}{2000000}$ under that load. A large balance constructed by Saxton, of Washington, and exhibited at the Paris Universal Exposition of 1867, turned on the addition of a weight of 1 mgr. when loaded with 20 kgrs. in each pan, a ratio of turning-weight to load of $\frac{1}{20000000}$, and a large American balance in the *Conservatoire des Arts et Métiers* at Paris, detects a variation of $\frac{1}{2}$ mgr. when loaded with 25 kgrs.[1]

185. Construction of delicate Balances. Fig. 83 represents a delicate chemical balance. The beam *AB* is of brass, and of such a form as to combine strength with lightness. The fulcrum, *F*, is a knife-edge of hardened steel, resting upon a horizontal plate of agate firmly fixed to the column *CD*, sustaining the balance. A long index attached to *AB* and moving before a

[1] For a table giving the sensibility of the principal balances exhibited at the Exposition of 1867, see *Report of U. S. Commissioners*, Vol. III., *Industrial Arts*, by F. A. P. Barnard; p. 485.

graduated scale G, serves to detect the slightest inclination of the beam. The pans P, P', are suspended from delicate knife-edges, A, B, and so adjusted as to bring the centre of gravity of each pan directly beneath the point of support. The beam is also furnished with a vertical adjusting screw, E, by means of which the centre of gravity can be brought as near to the fulcrum as is desired, thus regulating the sensibility according to the delicacy desired in any particular case. To prevent unnecessary wear of the knife-edge forming the fulcrum, the balance is arranged so that by turning the screw, O, the beam is lifted, and the fulcrum no longer rests upon the agate plate, in which position it is allowed to remain when not in use. A similar device is sometimes adapted to the wedges supporting the scale-pans. To screen the instrument from currents of air it is customary to inclose it in a glass case which is furnished with leveling-screws V, V', in order that the fulcum upon which the beam rests may always be brought to a horizontal position.

186. Construction of Weights and Method of Weighing. The large weights used with delicate balances (1 gramme and upwards) are generally made of brass, those of lesser denominations of sheet platinum. The brass weights are carefully turned and adjusted by filing; the smaller ones are made by cutting out a rectangular sheet of platinum of a known weight, as 1 gramme, and dividing this into equal parts, 10 for decigramme weights, 100 for centigrammes, and so on. Occasionally platinum wire is used instead of sheet platinum. As very small weights are inconvenient to use, milligrammes and fractions of a milligramme are frequently estimated by a different method. One arm of the scale-beam is divided into a scale of equal parts, and a small U-shaped bit of wire of known weight, called a *rider*, is placed upon it. The rider balances a greater or less weight in the opposite pan according to its position on the beam. The beam is readily graduated as follows. The rider weighing 1 cgr. is placed upon it and moved until it just balances a half centigramme weight, placed in the opposite pan. This point is marked, the weight removed from the scale pan, and replaced by a milligramme weight. The rider is now moved towards the fulcrum, until equilibrium is restored, and this point is also marked. Evidently when at the latter mark, the rider has the same effect as a weight of 1 mgr. in the adjacent pan, and when at the first made mark it corresponds to a weight of 5 mgr. If now the space between these be divided into 40 equal parts, and the division continued towards the extremity of the beam, each division will correspond to $\frac{1}{10}$ mgr.

The *Arc of Precision* of Gallois has recently been substituted for the method of weighing by the rider, by several of the best French and German makers. "This is an expedient for making delicate determinations of fractional weights, by deflecting more or less to the right or left an index needle attached to the beam of the balance beneath the centre of gravity. When this index points directly downward like the ordinary fixed needle of the balance, its effects upon the equilibrium of the balance is of course zero. When deflected so as to form an oblique angle with the horizontal axis of the beam, it contributes a portion of its weight, dependant on the amount of deflection, to the side toward which it is inclined. When, in the process of weighing, a position of the needle has been found which produces equilibrium, the fractional weight contributed by the needle is read upon a circular arc, which is situated immediately behind it, and suitably divided. The division, of course, must be made experimentally at the time of the construction the balance."[1] This method is more rapid than the ordinary process. Which is the more accurate, remains to be determined by experience.

187. Weighing with a False Balance. If one of the arms of a balance is longer than the other, the weights required in each pan to produce equilibrium will be inversely proportional to the length of the adjacent arm (Eq. 101, p. 88). A given weight in one pan may therefore be made to counterpoise a greater weight in the other by simply lengthening the arm of the balance to which it is applied. The dishonest dealer is thus furnished with a ready method of defrauding his customers, by placing the goods to be sold in the pan suspended from the longer arm. Such a deception may be detected by transferring the goods to the opposite pan, and replacing them by weights which previously balanced them, in which case the equilibrium will no longer be sustained. It is possible, however, to weigh truly with such a false balance. Let the article be weighed successively in each pan, the apparent weights thus found being P, P', respectively. Call x the true weight, and L, L', the arms of the beam. Then

$$L : L' :: x : P, \text{ and } L : L' :: P' : x;$$

whence $$x : P :: P' : x, \text{ or } x = \sqrt{PP'}. \quad (112.)$$

The true weight is therefore a mean proportional between the apparent weights.

188. Method of Double Weighing. Another means of accomplishing the same end is the following. The body to be weighed is placed in either of the scale-pans, and accurately balanced by means of fine shot or sand. It is then removed, and weights substituted for it until equilibrium is made with the shot or sand used as a counterpoise. The weights thus added evidently equal the weight of the body, as both have been applied at the extremity of the same arm to counterpoise a constant weight. As no balance can be made with absolutely equal arms, it is advisable to use the present method whenever great accuracy is required.

The process which we have just explained is commonly known as *Borda's Method of Double Weighing*, after the physicist of that name who has generally received the credit of having invented it. That distinction, however, properly belongs to Père Amiot, who made use of it some time previous to Borda.

189. Roberval's Balance. Fig. 84 represents a section of a form of balance often used where only a moderate degree of accuracy is required. The pans are attached to the vertical bars AC, BC' which are fastened by pivots to the extremities of the horizontal beams AB, $A'B'$, so that whether the pans rise or fall, AC and BC' remain vertical. An index I serves to show when equilibrium is attained.

In this apparatus it is indifferent on what part of the pan the weights are placed. For let P, P' be the weights placed so as to act at different distances from the fulcrums F, F'. Imagine two equal and opposite forces, P'', P''', each equal to P, applied at C, the centre of the pan on which P rests. Also two equal and opposite forces P^{iv}, P^{v}, each equal to P' applied at C'. The addition of these does not alter the equilibrium of P, P'. The balance is now acted upon by two equal forces P'', P^{iv}, applied at C, C',

[1] *Report of U. S. Commrs. to Paris Universal Exposition of* 1865; Vol. III., p. 484.

which balance each other, and also by the couples $P - P'''$, $P' - P^{v}$, which tend to produce rotation of the whole system of levers. The effect of these couples is balanced by the resistance of the pivots, hence the final result is the same as if P, P' were transferred to C, C', the centres of the scale-pans.

190. Steelyard. Another instrument adapted to the weighing of articles of moderate size is the common steelyard, the use of which dates back to the time of the ancient Romans. It consists of an iron bar AC, Fig. 85, near one extremity of which are two hooks resting upon pivots at A, B. The instrument is suspended by the hook H, and the article to be weighed, W, is hung upon the other hook. W is balanced by a small counterpoise P, which is movable along the beam. The greater the distance BD, the greater will be the weight balanced by P. The beam is furnished with a scale which is graduated in the following manner. If the hook B is placed at the centre of gravity of the beam, the graduation is effected by the process described in the explanation of the use of the *rider* (§ 182 p. 94) More commonly, however, the centre of gravity is at some point as G, beyond B. In this case, let E be a point from which the counterpoise must be suspended to just balance the beam when no weight is suspended from A. E is the zero of the scale. For calling w the weight of the beam, $w \times GB = P \times EB$. Let a weight W, be applied at A, and suppose P to balance this when at M. Then,

$$W \times AB = w \times BG + P \times BM = P \times EB + P \times BM = P \times EM.$$

If now a weight, W', be suspended from A, and D be the position of P when W' is counterpoised by it,

$$W' \times AB = P \times EB + P \times BD = P \times ED.$$

In these equations AB and P are constant, whence

$$W : W' :: EM : ED, \quad (113.)$$

Or the weight balanced by P is proportional to the distance of P from E. Hence if W be taken as 1 kgr., and W' as 20 kgrs., and the space ED divided into 20 equal parts, each of these will correspond to 1 kgr.

When P is placed at the extremity of the arm AC, it denotes the maximum weight which the instrument is capable of measuring, but the same beam may be used to determine greater weights by using a heavier counterpoise, or by suspending the weight W from a pivot nearer B, in which case a corresponding graduation is made on the other side of BC.

In the *Danish Balance*, Fig. 86, the counterpoise is fixed at one end of the beam, the article to be weighed suspended from the other end, and the beam is balanced upon a movable knife-edge fulcrum, from the position of which when equilibrium occurs, the suspended weight is determined.

191. Bent Lever Balance. This form of balance consists of a bent lever ABC, Fig. 87, moving upon a pivot at B. At one end is a hook A, to which a scale-pan is attached, and at the other a fixed counterpoise C. Let us denote the total weight of the beam and pan by w, their common centre of gravity being at G. On placing a weight W in the pan, its effect is to raise the arm BC. But as BC rises, the leverage of C is increased, while the leverage of W is at the same time diminished. Hence the arm BC ascends until $w \times BM = W \times BA$. A circular scale PR is attached to the balance. It is best graduated experimentally by placing weights of 1, 2, 3 kgrs. in the pan successively, and marking the point denoted by the index with corresponding numbers.

192. Platform Balance. The various forms of machines used for weighing bodies of considerable magnitude depend upon the principle of the compound lever. Fig. 88 shows the construction of one of the most common forms. It consists of two V-shaped levers AIB, CID, resting upon fulcrums A, B, C, D, fastened at I to a fourth lever FE. A vertical rod EH connects FE with the balance-beam HG. MN is a platform on which the article to be weighed is placed. This platform is furnished with four steel legs, $a'\ b'\ c'\ d'$, which when in place rest upon the levers AI, BI, CI, DI, at the points a, b, c, d. The pressure upon the platform is communicated to EF by means of the four lower levers, and thence by the vertical connecting-rod to GH, where it is balanced by a movable weight P, as in the steelyard. The V-shaped levers are used instead of connecting the platform directly with I, in order to render the effect of the mass weighed the same on whatever part of the platform it may rest. To compute the ratio of the pressure W, on the platform to the counterpoise G, we may consider the whole pressure as applied to either one of the four arms AI, BI, CI or DI. Suppose it to be applied at a. Then by the principle of the compound lever (Eq. 105, p. 89), the power multiplied by the continued product of the power-arms is equal to the weight multiplied by the continued product of the weight-arms. That is, $P \times GF' \times EF \times AI = W \times HF' \times IF \times Aa$. Thus if the length of $AI = 20$, $Aa = 4$, $EF = 20$, $IF = 5$, $GF' = 10$, $HF' = 1$, $G \times 10 \times 20 \times 10 = W \times 1 \times 5 \times 4$; whence $2000G = 20W$, or $W = 100G$. A weight of 1 kilogramme at G will therefore sustain a mass of 100 kilogrammes upon the platform. An additional counterpoise P, is also furnished, the arm HG being graduated in the manner described in the case of the steelyard.

II. *Wheel and Axle.*

193. Wheel and Axle. — Laws of Equilibrium. The *wheel and axle* consists of a cylinder or axle turning on gudgeons, to which is attached a larger cylinder, called the wheel, as shown in section in Fig. 89. The power is applied at the circumference of the wheel, the weight at the circumference of the axle, in such a manner that they tend to produce rotation in opposite directions. The power and weight are generally parallel, and applied along tangents to the sections of the wheel and of the axle respectively.

To ascertain the law of equilibrium, it is to be recollected that the moment of the power and weight relatively to the axis through C, must be equal. Hence calling P the power, W the weight, R the radius of the wheel, r the radius of the axle,

$$P \times AC = W \times BC, \text{ or } PR = Wr;$$

$$\text{whence } P : W :: r : R \quad (114.);$$

that is, there is equilibrium with the wheel and axle when

power : *weight* :: *radius of axle* : *radius of wheel.*

From (114.) it follows that $W = P\frac{R}{r}$. (115.)

If the power does not act tangentially, but has an oblique direction, as AP, Fig. 90, we have for equilibrium, $P \times CD = W \times r$, or calling a the angle made by AP with the tangent AD, $P \times R \cos a = W \times r$.

When the power and weight are applied by means of a rope the half-thickness of the rope must be added to the radii R and r to obtain a correct theoretical result. In this case, calling t the thickness, we should have $P(R + \frac{1}{2}t) = w(r + \frac{1}{2}t)$. (116.) Practically, however, this correction is unnecessary, as it is less than the deviation from theory produced by friction, rigidity of cordage, etc. The total pressure on the pivots sustaining the machine $= P + W$, if these forces are parallel.

194. The chief advantage of the wheel and axle lies in the power it gives us of continuing the motion which raises the weight, so that with a machine of moderate size we can raise a heavy weight through any required distance. Practical applications are seen in the common windlass, crane and capstan.

The wheel and axle may evidently be considered as a lever, of which R, r are the arms.

195. Chinese Capstan. Since the weight is greater than the power in the ratio of the radius of the wheel to the radius of the axle, the weight sustained by a given force may be increased, either by increasing the size of the wheel, or diminishing the size of the axle. Practically, however, it is inconvenient to use a very large wheel, while a small axle does not possess the necessary strength. In the machine known as the Chinese Capstan, Fig. 91, this difficulty is overcome in the following manner. The axle is made in two parts, D and C, one of which has a greater diameter than the other. The rope sustaining the weight W is wound about this axle, so that its two sections, ED, FC, tend to produce motion about the axis AP in different directions. To find the conditions of equilibrium, call r the radius of the larger axle, r' that of the smaller one. The weight W is sustained by the cords ED, FC, each of which therefore exerts a force $\frac{1}{2}W$ tangential to the axis about which it is wound. Hence that portion of the weight sustained by CF tends to produce motion in the direction of the arrow, having a moment $= \frac{1}{2}Wr'$, and acting in the same direction as the power P, while the portion sustained by ED tends to produce motion in an opposite direction, having a moment $= \frac{1}{2}Wr$. Hence equilibrium ensues when $PR + \frac{1}{2}Wr' = \frac{1}{2}Wr$, or $PR = \frac{1}{2}W(r - r')$, whence $P : W :: \frac{1}{2}(r - r') : R$. (117.) The efficiency of this machine therefore depends not upon the absolute size of the axles, but merely upon the difference between their radii, so that they may be made as large as is necessary to bear the strain put upon them without diminishing the mechanical advantage.

196. Combined Wheels and Axles. It is often necessary to combine wheels and axles, and when the object is to gain an increase of power, the method most commonly adopted is that of toothed wheels. The teeth of the wheels which gear into each other being made of suitable form, motion is communicated by the pressure of the teeth of the first wheel, or *driver*, as it is called, upon those of the second wheel, or *follower*.

The conditions of equilibrium may be found thus. Let AB, $A'B'$, Fig. 92, be two cogged-wheels having their centres at C, C'. Call R, R', the radii of the wheels, r, r' the radii of their axles, upon which P and W act. If P and W are supposed to be in equilibrium, the force at D communicated from P by means of the wheel AB, and tending to produce a right-handed rotation in $A'B'$, must be balanced by an equal force acting in an opposite direction, communicated from W by means of the wheel $A'B'$, and tending to produce a right-handed rotation in AB. Calling this force F, then in the wheel AB, $FR = Pr$, and in $A'B'$, $FR' = Wr'$. To estimate the effect of the cogged-wheels alone, suppose the axles to have equal radii, that is $r' = r$. Then from the preceding equations,

$P : R :: F : r$, and $W : R' :: F : r$, whence $P : W :: R : R'$ (118.), that is, *power is to weight as the radius of the cogged-wheel on which P acts, is to the radius of the wheel acted upon by W.* As the circumferences of AB, $A'B'$, are proportional to their radii, proportion (118) may be written $P : W ::$ *circumference of* AB : *circumference of* $A'B'$; or since the teeth are equal in size, the number of them contained in the circumferences of AB, $A'B'$, are proportional to those circumferences, and we may write the equation of equilibrium as follows, denoting by n, n', the number of teeth in the driver and follower respectively, $P : W :: n : n'$. (119.)

197. System of any Number of Wheels. To find the conditions of equilibrium in a system of any number of wheels in which power is transmitted by means of cogs, as in Fig. 93, in which the teeth of each wheel are put in motion by the teeth of a pinion attached to the preceding wheel,[1] we proceed as follows: Let R, R' R'', be the radii of the wheels, r, r', r'', the radii of the corresponding pinions. Denoting by F, F', F'', etc., the powers applied by the pinions at the circumference of the wheels AB, $A'B'$, $A''B''$, we have, by the course of reasoning applied in the preceding paragraph, $PR = Fr$, $FR' = F'r'$, $F'R'' = Wr'$. Hence

$$W = F'\frac{R''}{r''} = F\frac{R'}{r'} \cdot \frac{R''}{r''} = P\frac{R}{r} \cdot \frac{R'}{r'} \cdot \frac{R''}{r''}, \text{ and}$$

$$P : W :: r \times r' \times r'' : R \times R' \times R''. \quad (120.)$$

That is, *power is to weight as the continued product of the radii of the pinions is to the continued product of the radii of the wheels.*

198. Spur, Crown and Bevel Wheels. A wheel in which the teeth are raised upon the circumference as in Fig. 93, is called a *spur-wheel.* If the teeth are raised upon the side of the wheel, and parallel to the axis as in Fig. 94, it is a *crown-wheel*, and if they are raised upon a surface inclined to the plane of the wheel (Fig. 95), it is known as a *bevelled-wheel.* An inspection of the figures will show that *spur-wheels* communicate motion in their own planes, while a crown-wheel engaging with a spur pinion (Fig. 94), causes a motion of the latter about an axis at right angles to its own. Bevelled wheels are formed of frustra of cones, and communicate motion from one axis to another, inclined at any angle to it. When a pinion works in a straight bar, having teeth raised upon it, the combination is known as a *rack and pinion.*

199. Form of Teeth. In combinations of circular wheels, the teeth should be so cut that a perfectly uniform motion may be transmitted from one wheel to the other. Thus the two cog-wheels represented in Fig. 96, should move with the same relative velocities that would ensue if the two circles AB, $A'B'$, called the *pitch-circles* of the wheels were made to roll upon each other. That this may be the case, it is necessary that MS, the perpendicular to the surfaces of contact of the teeth at any moment of the revolution shall pass through D, the point of contact of the pitch-circles. To demonstrate this, suppose the circle AB to roll upon AB'. The actual linear velocity of any point as D in the circumferences of the circles, must be the same in each, by the conditions of the motion. Denote by a the angle through which AB revolves in an element of time, and by a' that passed through by $A'B$; a and a' are called the *angular velocities* of AB, $A'B'$. The lengths of the arcs DK, DK' subtending these angles, must be equal. Now as *arc* = *angle* $\times$ *radius*, $DK = CD \times a$, $DK' = C'D \times a$, whence $CD \times a = C'D \times a'$ or $a : a' :: C'D : CD$. (121.)

[1] When teeth are cut upon an axle it is called a pinion.

Suppose now that the wheels are furnished with teeth. To fulfil the required condition, proportion (121) must still hold. Consider the wheels as revolving, and let M be the point of contact of two teeth. The pressure of the driving tooth upon the following one causes motion, the teeth being in what is known as *sliding-contact*. By the conditions of the motion, the two teeth must at the moment considered, have equal velocities in the direction MS. The space passed over by the tooth K in an element of time $= CM \times a$, that passed over by the tooth $K' = C'N \times a'$, and as these spaces are equal $a : a' :: C'N : CM$ (122). If the values of the angular velocities in the case of motion communicated by teeth are supposed equal to those obtained by the rolling of the pitch-circles upon each other, a comparison of proportions (121) (122), gives $C'D : CD :: C'N : CM$, (123), a condition which can hold only when the triangles CDM, $C'DN$ are similar, in which case MN must pass through D, the point of contact of the pitch-circles.

200. In all well-constructed gearing the teeth of the wheels are so cut as to fulfil this condition. Great ingenuity has been spent in devising the forms of teeth suitable in different cases. For the details of such investigations, the student is referred to any of the standard treatises on Mechanism. It may be stated in general terms that it follows from the preceding demonstration, that the teeth should have the form of *epicycloids*, generated by the rolling of a circle of diameter determined by the size of the teeth, upon the pitch-circles of the wheels.

III. *Cord.*

201. Case of Forces applied at different points of a Cord. *The forces acting at various points of a cord are in equilibrium when they bear the same relations to each other in magnitude and direction, as if they were in equilibrium about a single point.*

Let us first take the case of three forces. AB, Fig. 97, is a rope, at one extremity of which are applied forces, P, P', and at the other a force P''. To find their relations when in equilibrium, lay off AP, AP', representing P, P' respectively, and complete the parallelogram $APRP'$. R, represented by AR, is their resultant. Now the rope being flexible it must assume such a direction that the directions of R and the remaining force P'' shall be in the same straight line, RP''. Hence the force P'' required to balance R, must be equal to R, and the forces P, P', P'', are represented in magnitude and direction by the three sides, AP, PR, RA, of the triangle APR, in which case they would also be in equilibrium if applied at a single point.

The same theorem holds for any number of forces. Let $ABCD$, Fig. 98, be a cord with forces $P_1, P_2, P_3, P_4, P_5, P_6, P_7, P_8$, applied as there shown, and represented in magnitude and direction by the lines AP_1, AP_2, BP_3, BP_4, etc. The resultant of P_1, P_2 (R_1), takes the direction AR_1, which must lie in the prolongation of BA, and hence may be transferred to B without alteration of its efficiency (§ 105, p. 58). In like manner, the resultant R_2 of R_1, P_3, P_4, must lie in the prolongation of CB, and hence may be transferred to C. The resultant R_3 of R_2, P_5, P_6, must lie in the prolongation of DC, and may be transferred to D without alteration of the equilibrium. If, then, there is equilibrium, the resultant R_3 arising from the combination of P_1, P_2, P_3, P_4, P_5, P_6, must be equal and opposite the resultant, S, of P_7, P_8, which would also be the case if all the forces under consideration were applied at a single point.

In this demonstration the cords are supposed to be perfectly flexible and destitute of weight.

202. Illustrations. The cord is extremely useful in changing the direction of the motion, but as the tension is the same throughout its whole extent, there is no mechanical advantage gained. Most interesting examples of the employment of the cord for this purpose, occur among the muscles of the human body. The tendon of the muscle raising the eyelid (Fig. 99,) runs through a loop A, so that the contraction of the muscle in the direction BA, moves the lid in the direction CA. Similar examples may be seen in the tendons of the fingers and toes.

203. Weight Supported by Segments of a Cord. If a weight W be attached to a cord in the manner represented in Fig. 100, the forces TT' which must be exerted at EF, to sustain it, may be found from equations (27), (28), $T = W\frac{\sin DAC}{\sin CBA}$, $T' = W\frac{\sin BAC}{\sin CBA}$. If we suppose $BAC = DAC$, $T' - T$. In this case, denoting BAD by $2a$, $T = T' = W\frac{\sin a}{\sin 2a} = W\frac{\sin a}{2 \sin a \cos a} = \frac{W}{2 \cos a}$. (124.) The magnitude of the forces T, T', which measure the tension of the rope, is evidently inversely proportional to the cosine of half the angle made by the segments of the rope ($\cos a$). If these segments are parallel, in which case $a = 0$, $T = T' = \frac{1}{2}W$.

That the segments of the cord may be in the same straight line, that is, that the rope may become horizontal, the angle BAD must $= 180°$. That is, $a = 90°$, in which case $T = T' = \frac{W}{2 \cos a} = \frac{W}{0} = \infty$. That is, *it requires an infinite power to cause loaded cord to become horizontal.* Still further, as the whole weight of any material cord may be considered to be concentrated in its centre of gravity, and to act vertically downward through that point, *it is impossible to stretch any material cord into a horizontal position.*[1]

It is also evident from the preceding demonstration, that *the smallest power can be so applied as to produce an unlimited amount of force.*

204. Funicular Polygon. The principles of the transmission of pressure by a cord, determine the position taken by a rope when acted upon by a number of forces applied at different points, as represented in Fig. 101. The figure $ABCDEF$, formed by the segments of the rope, is known as the *funicular polygon.* A similar figure is assumed by a series of beams united at their extremities by pivots and suspended as shown in Fig. 102. The forces acting upon the polygon in this case, are evidently the weights of

[1] Compare case of *Toggle Joint*, § 91 p. 53.

the separate pieces AB, BC, CD, DE, which may be considered as concentrated at their respective centres of gravity. It is found that if the funicular polygon be inverted (as shown by the dotted lines of the figure), it is the strongest form in which a given number of beams can be united to resist pressure, a fact of importance in the art of construction.

205. Catenary. If the number of the pieces AB, BC, etc., be increased indefinitely the funicular polygon has an infinite number of sides, and becomes a curve (Fig. 103), which is known as the *catenary*. An unloaded cord, or a chain composed of a great many links, when suspended by both ends, is an example of this curve. A knowledge of the properties of the catenary is of the greatest value in the construction of suspension bridges, which are simply bridges suspended from two parallel chains, (catenaries) stretched across a river. If the position of the extremities AB, the total length of the catenary ACB, and its weight are known, it is possible from analytical construction, to determine the tension at any point of the curve, both when unloaded and when loaded with a given weight, so that the strength necessary to be given to the chain-cables of such a bridge to enable them to resist the strain brought to bear upon them, can easily be calculated.

The inverted catenary is used in the construction of domes and arches of masonry.

IV. *Pulley.*

206. Pulley. The *pulley* consists of a circular disc, AB, Fig. 104, moving upon an axis at C, and having a cord or band to which the power and weight are applied, laid in a groove running over its circumference. The axle supporting the machine may be fixed or otherwise. In the first case, the pulley is said to be *fixed*, Fig. 104, in the second *movable*, Fig. 105.

207. Conditions of Equilibrium in Fixed Pulley. The power being applied at P, and the weight at W, these tend to rotate the pulley in opposite directions; hence, for equilibrium, their moments must be equal; that is, $W \times CB = P \times AC$, or as $AC = CB$, both being radii of the same circle, $W = P$. (125.) Hence, *with a fixed pulley there is equilibrium when the power applied is equal to the weight.*

208. Conditions of Equilibrium in Movable Pulley. *Case I. When the Segments of the Cord are parallel.* Fig. 105 represents such a pulley, with the power P applied at A, and the weight W suspended from the axis C. W and P each tend to rotate the pulley about B, but in opposite directions, and for equilibrium their moments must be equal; that is, $W \times BC = P \times AB$. But $AB = 2BC$, as AB is the diameter of the pulley, hence $W \times BC = P \times 2BC$, or $W = 2P$. (126.) That is, with a single movable pulley, if the segments of the cord are parallel, *there is equilibrium when the weight equals twice the power.*

Case II. When the Segments of the Cord are not parallel. Fig. 106 represents a pulley in which the segments of the cord are oblique. Suppose the power P to be applied along the line PA, and W to be suspended from C, and denote the angle PNH, made by the segments of the cord by $2a$. W and P tend to rotate the pulley in opposite directions about B, the first point of contact of the cord with the pulley. The moment of $W = W \times BM$, that of $P = P \times BD$, BD being drawn from B perpendicular to PA produced. Hence, for equilibrium, $W \times BM = P \times BD$. But $BM = BN \sin BNM = BN \sin a$,[1] and $BD = BN \sin BND = BN \sin ANB = BN \sin 2a$. Hence $W \times BN \sin a = P \times BN \sin 2a$, or $W \sin a = P \sin 2a$, whence $W \sin a = P\ 2 \sin a \cos a$, and $W = 2\ P \cos a$. (127.) That is, if the segments of the cord are oblique, *there is equilibrium when the weight equals twice the power into the cosine of half the angle made by the segments.*

If the segments become parallel, in which case $2\ a = 0°$, and $\cos a = 1, W = 2\ P$, as demonstrated under *Case I.* That the segments may lie in the same straight line, $2\ a$ must become 180°, in which case, $\cos a = 0$ and $P = \frac{W}{0} = \infty$, as in the case of the cord under similar conditions.

209. Pressure on Support. With a fixed pulley in which the segments of the cord are parallel, the pressure on the axis of the pulley evidently $= P + W$. With a movable pulley and parallel segments of cord, the pressure on the hook $H = \frac{1}{2}\ W = P$, as the weight is sustained by the two segments AP, BH. When the segments make an oblique angle, the pressure exerted upon the hook in any given direction may be calculated by means of the triangle of forces.

210. Combinations of Pulleys.—Block and Tackle. Combinations of pulleys are employed in the common *block and tackle*, and in various other forms of hoisting machines. There are several ways of arranging the pulleys composing the system, the most important of which we proceed to consider. Two common forms of hoisting pulleys, known as the block and tackle, are shown in Figs. 107, 108, in which systems the same rope runs over all the pulleys. In both figures A, B, are fixed, and C, D, movable pulleys. Since the weight W is supported by the united action of the segments of the cords, AC, CB, BD, DE, each of these must sustain an equal strain. If there are n segments, each will bear $\frac{1}{n}$th of the whole weight, and as the tension of the cord is the same throughout its whole extent, the power P required to balance W is $\frac{W}{n}$. Hence $W = n\ P$ (128.) That is, *in any combination of pulleys in which one rope runs through the whole system, the weight is equal to the power multiplied by the number of segments of the cord.*

The downward pressure at $H = W + P = n\ P + P = P\ (n + 1.)$ (129).

211. Geometrical Pulley. The system represented in Fig. 109, is known as the *Geometrical Pulley.* In it there are as many separate ropes as there are movable pulleys. The relation of weight to power may be

[1] P is supposed to be so applied that HNC, CNP are equal.

found as follows. If a power P be applied to the pulley A by means of a rope running over a fixed pulley S, we find from the law of the movable pulley, that the upward force exerted by A upon $B = 2\ P = P \times 2$. Since this force $2\ P$ becomes a new power applied to the movable pulley C, we have *Upward Force exerted by B upon* $C = 4\ P = P \times 2^2$. In like manner. *Upward Force exerted by C upon* $D = 8\ P = P \times 2^3$. Hence the weight which D can uphold $= 16\ P = P \times 2^4$. In the same manner for any number of pulleys as n, we should find that $W = P \times 2^n$, (130), so that with this arrangement *the weight is equal to the power exerted, multiplied by 2 raised to that power whose exponent is the number of movable pulleys.*

The pressures upon the hooks are as follows:—

Pressure on $N = \frac{1}{2}\ W = \frac{1}{2}\ P \times 2^n = P \times 2^{n-1}$

Pressure on $M = \frac{1}{2}$ *Pressure exerted by D on C* $= \frac{1}{2}$ *Pressure on* $N = \frac{1}{2}\ P \times 2^{n-1} = P + 2^{n-2}$

Pressure on $L = \frac{1}{2}$ *Pressure exerted by C on B* $= \frac{1}{2}$ *Pressure on* $M = \frac{1}{2}\ P \times 2^{n-2} = P \times 2^{n-3}$

Pressure on $H = \frac{1}{2}$ *Pressure exerted by B on A* $= \frac{1}{2}$ *Pressure on* $L = \frac{1}{2}\ P \times 2^{n-3} = P \times 2^{n-4}$

212. Another system of pulleys is shown in Fig. 110. The power applied at P produces by means of the rope PAH an upward pull at $H = P$. The rope ABG exerts an upward pull at $G = 2\ P$, the rope BCF exerts an upward pull at $F = 4\ P$, and the rope CDE exerts an upward pull at $E = 8\ P$. Hence, the total force exerted on $W, = P + 2\ P + 4\ P + 8\ P = 15\ P$ or $W, = P \times 2^{n-1}$ (131), if $n =$ number of movable pulleys.

The pressure upon the hook $K = P + W = P + P\ (2^{n-1}) = 2^n\ P$.

213. The relation of weight to power in the various combinations of these systems frequently occurring in practice may easily be calculated by a process similar to that by which the preceding formulæ were obtained. For example, let us take the arrangement of pulleys shown in Fig. 111. The power P is applied to a cord running over the pulley A, which alone would cause an upward force on $B = 2\ P$. The tension of the cord running from A over the fixed pulley $D = 2\ P$. This force is also communicated to B, as the cord is fastened to the axis of that pulley, causing the total upward pull exerted by B to be $4\ P$. Hence in this case $W = 4\ P$.

In a similar manner it may be shown that with the combination represented in Fig. 112, $W = 5\ P$.

214. Differential Pulley. The liability of a system consisting of a large number of pulleys to get out of order, and the great length of rope required for such combinations, together with a number of other practical difficulties, render them undesirable for ordinary use. The device known as the *Differential Pulley* (Fig.113), remedies some of these difficulties. Its principle is similar to that of *the Chinese Capstan* described on p. 98. The upper pulley, AE is fixed, the lower one FG, is movable. AE is composed of two pieces virtually forming two pulleys EB, AD, firmly connected, and moving about a common centre C. An endless chain $BPDAGE$ runs over both pulleys as indicated by the arrows in the figure. The weight is hung from the axis of the movable pulley. To find the conditions of equilibrium in this machine, call P the power, W the weight, r the radius of EB, r' the radius of AC. Then the fixed pulley is acted upon by three forces as follows. P, acting with an arm CB (r), and $\frac{1}{2}\ W$ with an arm AC (r'), both of which tend to produce a right-handed rotation; and $\frac{1}{2}\ W$ with an arm EC (R), tending to produce a left-handed rotation. For equilibrium the opposite moments of these forces must be equal, that is,

$Pr + \frac{1}{2} Wr' = \frac{1}{2} Wr$, or $Pr = \frac{1}{2} W(r - r')$, whence $P : W :: \frac{1}{2}(r - r') : r$. (132). The efficiency of this form of pulley therefore depends simply upon the difference between the radii of the two portions of the fixed pulley.

The friction of the differential pulley is enormous, but a practical advantage is gained thereby, from the fact that it will not *run down* when the power P is removed, the weight w not being sufficient to overcome this resistance.

215. Effect of Weight of Pulleys. In the foregoing discussions we have not taken the weight of the pulleys into account, which is necessary in practice, as they are frequently very heavy. With systems of movable pulleys this weight generally acts in opposition to the power, rendering a considerable portion of the applied force practically useless. A still greater loss of power occurs because of the friction and rigidity of the cordage employed, so that a very large percentage of the power is lost. The different velocities with which the various pulleys composing a system move, is also a source of practical difficulty. For these reasons complicated systems of movable pulleys are rarely used. The greatest value of the pulley is to change the direction in which the power acts.

V. *Inclined Plane.*

216. Definition; Conditions of Equilibrium. The inclined plane is considered as a perfectly hard and inflexible plane, inclined at an angle to the horizon.

To find the conditions of equilibrium, suppose W to be the weight of a body resting upon the inclined plane of which MN is the vertical projection. Let the power balancing it be denoted by F, acting in the line AF,[1] and making an angle β with the inclined plane. Denote the inclination of MN by α. The weight, W, of the body acts vertically downward along the line AW. Let AW represent the magnitude of this force, and resolve it into two components, P and R, one of which is perpendicular to the plane MN, and the other parallel and opposite to the force F. These components are represented by AP, PW, respectively. The total force of the component P is exerted in producing a pressure upon the inclined plane MN, to which it is perpendicular, and is wholly balanced by the reaction produced; hence it is merely necessary that the remaining component R shall be balanced by the power F, in order that the weight may be supported; that is, for equilibrium, the force F must be equal to the component R. If this is the case, the three sides, AW, WP, PA, of the triangle AWP, taken in order, will represent the three forces, W, F, P, under whose united action the weight is sustained. Hence, for equilibrium, $F : W :: WP : WA :: \sin WAP : \sin APW$.
But $WAP = \alpha$, $APW = 90° + \beta$, hence

[1]We confine our attention to the case in which the force F lies in the plane MNO, at right-angles both to the inclined plane and to the horizon.

$$F : W :: \sin\alpha : \sin 90^\circ + \beta, \text{ or } F : W :: \sin\alpha : \cos\beta. \quad (133.)$$

That is, *with the inclined plane there is equilibrium when power is to weight as the sine of the angle of inclination of the plane is to the cosine of the angle made by the power with the plane.*

217. Pressure on Plane. The pressure on the plane, P, is represented by AP, and may be determined from the proportion

$$P: W :: AP : AW :: \sin AWP : \sin APW ::$$
$$\sin(90^\circ - \alpha - \beta) : \sin(90^\circ + \beta), \text{ or}$$
$$P : W :: \cos(\alpha + \beta) : \cos\beta \quad (134.),$$

a proportion from which the pressure on the plane may be computed in terms of the weight of the body considered.

By combining proportions (133), (134), we obtain the following:

$$P : F :: \cos(\alpha + \beta) : \sin\alpha. \quad (135.)$$

218. Special Cases. Proportion (133) is much simplified, when the power F acts either parallel to the plane, or horizontally. In the former case, FA and WP become parallel to MN, consequently $\beta = 0^\circ$. Therefore, (133), (134), (135), assume the form

$$F: W :: \sin\alpha : \cos 0^\circ :: \sin\alpha : 1. \quad (136.)$$
$$P: W :: \cos\alpha : \cos 0^\circ :: \cos\alpha : 1. \quad (137.)$$
$$P : F :: \cos\alpha : \sin\alpha :: 1 : \text{tang}\,\alpha. \quad (138.)$$

Denote the length of the inclined plane, MN, by L, the height, NO, by H, and the base, MO, by B. Then

$$\sin\alpha = \frac{H}{L}, \ \cos\alpha = \frac{B}{L}, \ \text{tang}\,\alpha = \frac{H}{B}.$$

By substituting these values for $\sin\alpha$, $\cos\alpha$, $\text{tang}\,\alpha$, in (136), (137), (138), the following proportions are obtained.

$$F : W :: \sin\alpha : 1 :: \frac{H}{L} : 1, \text{ whence } F : W :: H : L. \quad (139.)$$

$$P : W :: \cos\alpha : 1 :: \frac{B}{L} : 1, \text{ whence } P : W :: B : L. \quad (140.)$$

$$P : F :: 1 : \text{tang}\,\alpha :: 1 : \frac{H}{B}, \text{ whence } P : F :: B : H. \quad (141.)$$

That is, when the direction of the power is parallel to the plane,

I. *Power is to weight as the height of the plane is to its length.*
II. *Pressure on plane is to weight as base of plane is to its length.*
III. *Pressure on plane is to power as base of plane is to its height.*

If F is parallel to the base of the plane, $\beta = -\alpha$, in which case proportions (133), (134), (135), assume the following forms:

$$F : W :: \sin\alpha : \cos(-\alpha) :: \sin\alpha : \cos\alpha :: \frac{H}{L} : \frac{B}{L}, \text{ whence}$$
$$F : W :: H : B. \quad (142.)$$

$$P : W :: \cos 0^\circ : \cos\alpha :: 1 : \cos\alpha : 1 : \frac{B}{L}, \text{ whence}$$
$$P : W :: L : B. \quad (143.)$$

$$P : F :: \cos 0^\circ : \sin \alpha : 1 : \sin \alpha : 1 : \frac{H}{L}, \text{ whence}$$

$$P : F :: L : H. \quad (144.)$$

That is, when the direction of the power is parallel to the base of the plane,

I. *Power is to weight as height of plane is to base.*
II. *Pressure on plane is to weight as length of plane is to base.*
III. *Pressure on plane is to power as length of plane is to height.*

219. Direction of Power for Maximum Weight and Pressure. Solving (133) relatively to W, we have $W = F\frac{\cos \beta}{\sin \alpha}$, a quantity which for any given inclined plane is proportional to $\cos \beta$, and therefore has its maximum value when $\beta = 0^\circ$ and $\cos \beta = 1$. Hence *a given power will sustain the greatest weight when it acts parallel to the plane.*

The greatest pressure upon the plane produced when a given weight is resting in equilibrium upon it occurs when the power is parallel to the base of the plane. For from (135) $P = F\frac{\cos (\alpha + \beta)}{\sin \alpha}$, a quantity proportional to $\cos (\alpha + \beta)$, and which is greatest when $\cos (\alpha + \beta) = 1$, in which case $\beta = -\alpha$, and F is parallel to the base of the plane.

The maximum pressure is therefore $P = \frac{F}{\sin \alpha}$.

In the foregoing demonstrations, no account whatever has been taken of friction, an element which causes a great variation from the preceding results. Its effect will be discussed in a succeeding chapter. The mechanical advantage of the inclined plane lies in the fact that a portion of the weight is supported by the plane itself.

220. Conditions of Equilibrium of Bodies Resting on Contiguous Planes. Suppose two bodies, A, B, of weights W, W' respectively, to rest on two contiguous planes of the same height as shown in Fig. 115, B is connected with A by means of a cord passing over a pulley at N. It is required to find the relation of the weights W, W' that the system may be in equilibrium.

Call P the power parallel to NO required to sustain W, P' that parallel to MN required to sustain W', and denote NR by H, MN by L, NO by L'. As the power in both cases is parallel to the plane,

$$P : W :: H : L, \text{ whence } P = W\frac{H}{L}.$$

$$P' : W' :: H : L', \text{ whence } P' = W'\frac{H}{L'}.$$

But that W and W' may be in equilibrium P must be equal to P', in which case $W\frac{H}{L} = W'\frac{H}{L'}$, or $W : W' :: L : L'$, (145) that is, the *weights must be to each other as the lengths of the planes in which they rest.*

221. Practical Example. The best examples of inclined planes are seen on roads leading over hills, the power required to draw a vehicle over them increasing with their steepness. In passing over a level road, the horse has merely to overcome the friction of the wheels, but on an inclined plane he has in addition to lift a fraction of the load, the magnitude of which depends upon the steepness. Thus if the rise is 1 m. in 20 m., $\frac{1}{20}$ of the load must be lifted, for 1 m. in 40 m., $\frac{1}{40}$ of the load, and so on.

Interesting illustrations also occur in the various mountain railways which have recently been constructed both in this country and in Europe. The first of these in date of construction is the Mt. Washington Railroad, which was begun in May, 1866, and finished in July, 1869. The road starts from a point on the west side of the mountain, 813 m. above the level of the sea, and ascends with an average grade of 1 m. in 4 m. to the summit, which lies at an elevation, 1918 m., making the vertical height of the inclined plane equal to 1105 m. The grade being far too steep to be ascended by an ordinary locomotive, which acts solely by the friction of the wheels on the rails, a third rail is introduced, lying midway between the other two, and furnished with cogs. The middle rail is in fact a *rack.* A strong cogged-wheel driven by the locomotive works in this rack, and thus causes the locomotive and attached car to ascend. The steepest grade upon the road is 375 mm. per metre., or a little over 1 m. in 3 m. There are nine curves with radii varying from 151 m. to 288 m., and the road passes at one point over a trestle-work bridge 9.2 m. in height, and with an inclination of over 1 m. in 3 m.

The locomotive at present used weighs 5900 kgr. The carriages for passengers resemble a horse-car, and seat 48 persons. The locomotive is always below the car whether ascending or descending. The actual time of ascent is 90 minutes, a rate of over 3 km. per hour; the time of descent is somewhat shorter, and the descent is accomplished by means of gravity alone. The arrangements for controlling the motion are excellent, consisting of (1) a form of friction-brake attached to the driving-wheels, (2) the power of reversing the driving-wheels, and (3) an atmospheric brake attached to the cars. The road is kept open from the middle of June to the first of October, and is well patronized during the season of summer travel. The inventor of the peculiar form of locomotive and rails used in its construction is Mr. Sylvester Marsh, to whose energy the road owes its existence.

Another very remarkable mountain railway is that of Mt. Rhigi, in Switzerland, which was constructed by Riggenbach, upon the same general plan as the one at Mt. Washington. It was begun in Nov., 1869, and finished late in the summer of 1870. The vertical height of Mt. Rhigi above the sea is 1800 m., and above Lake Lucerne 1360 m. The road is 5340 m. in length, with a total ascent of a little over 1200 m., an average slope of over 220 mm. per metre. There are several curves, all of which have a radius of 180 m. At one part of its course the railroad passes through a tunnel 80 m. long, and shortly after emerging from this crosses a ravine 23 m. deep, upon a bridge 75 m. in length, and having an inclination of 250 mm. per metre. The weight of the locomotive when ready for use is about 12,500 kgr., being much larger than those used in this country. The velocity attained is 6.4 km. per hour, so that the ascent occupies only about that time. The passenger cars seat 54 persons, and when empty weigh 3970 kgrs. It is customary to have a man walk before the engine to clear the road of any obstructions that may occur.

The mountain railroad leading from Ostermundingen to connect with the railroad to Berne, is also of importance. It was constructed by Riggenbach. Unlike the two roads already mentioned, it was built, not for the purpose of carrying tourists to the mountain-summits, but to transport the building stones quarried at Ostermundingen to a place from which they could be taken by the ordinary railroads. It is composed of two nearly level portions united by a grade of 100 mm. per metre, which is half a kilometre in length. The total length of the road is a little over 2 km. The existence of both levels and inclines upon the road, rendered necessary a modification of the Rhigi locomotive, as the cogged-wheel and rail are unnecessary upon the levels. Hence the middle rail is made several centimetres higher than the exterior ones, and is omitted upon the level portions of the road, so that the cogged-wheel engaging in it has such an elevation above the ground, that it does not at all interfere with the working of the locomotive when upon the level.

The railroad up Mt. Cenis is perhaps more widely known than any of the preceding. The break in railroad connections (77 km.) from St. Michel to Susa led the government of Sardinia to undertake the construction of a tunnel through the mountain, in order to secure a continuous line of rail from France to Italy. The extreme length of the tunnel (12,220 m.) caused its progress to be very slow, and while the excavation was going on, Mr. J. B. Fell, an English engineer, proposed to construct a line over the mountain, using a peculiar system of rails and locomotive. Preliminary trials of the system were made in England, and a line was afterwards built up the side of Mt. Cenis. The length of this railroad is 1960 m., and its vertical ascent is 151 m. The line begins at a height of 1622 m., and terminates at a height of 1773 m. above the sea-level. Its mean grade is 1 m. in 13 m. (77 mm. per metre), and the maximum is 1 m. in 12 m. (83 mm. per metre). There are a number of curves with radii of from 40 m. to 300 m. The device for enabling the locomotive to climb the steep grade is somewhat different from those in use on the railroads at Mt. Washington and Mt. Rhigi. It consists of a middle rail, which is firmly embraced by two horizontal drivers, which furnish the necessary friction. There are also four vertical drivers of the ordinary form, which are used alone when the grade is less than 40 mm. per metre.

Another railway upon the Fell system is now in process of construction in Brazil. It is designed to convey the coffee raised in the elevated district of Cantagallo over the Organ Mountains to the lowlands. The length of the railway is about 12,500 m., and its gauge is 1.1 m., which is identical with that of the Mt. Cenis road. The average grade is about 77 mm. per metre, and the curves have a radius of 40 m. The centre rail will be used only on the steep inclines, which is also the case on the Mt. Cenis road.

Still another celebrated example of the mechanical power under consideration is the Timber Slide of Alpnach. This is a gigantic inclined plane about 14,000 m. in length, leading from the forests on the summit of Mt. Pilatus, one of the Swiss Alps, to the borders of the Lake of Lucerne. It was constructed in 1812, to convey the fir-trees from the mountain-forests to the lake, from which they could easily be carried to the Rhine. The mean inclination is 3° 14′ 20″, the greatest slope being 22° 30′. It is necessarily somewhat circuitous, owing to the unevenness of the ground. The trees, with their branches lopped off and divested of their bark, are placed in the trough of the slide, and descending by their own gravity, acquire a velocity so great that they reach the lower end in a very few minutes.

VI. *Wedge.*

222. Explanation of Action. If we suppose the power P, to be applied to ON, as represented in Fig. 116, pushing the inclined plane in the direction OM, and thus raising W, we have the mechanical power, known as the *wedge*. The relation between P and W may be estimated as follows: — Since the effect of a horizontal force in moving the plane is the same at whatever point of NO it may be applied, suppose it to act along PR, a horizontal line passing through A, the point of contact of MN with the vertical through the centre of gravity of the body resting upon it. Draw AB, BC, parallel to the directions of P and W, respectively, and complete the triangle ABC by drawing AC perpendicular to MN, as the reaction of the plane (R) assumes that direction. Then P, W and R, are represented in magnitude and direction by BC, AB, AC, respectively. Hence

$$P : W :: BC : AB :: NO : MO, \text{ or } P : W :: H : B. \quad (146.)$$

Also $P : R :: BC : AC :: NO : MN$, or $P : R :: H : L$. (147.)

That is, *power is to weight as the back of the wedge is to its base*, and *power is to reaction on face as the back of the wedge is to the length of its face.*

223. Common form of Wedge. The most common form of wedge consists of two inclined planes mounted base to base, as shown in Fig. 117. Suppose the power $2P$ to be applied at the middle of the back, and to act along the line DA, which bisects the angle BAC. The advance of the wedge is opposed by pressures R, R', at right-angles to the faces AB, AC, and if there is equilibrium the directions of $2P$, R, R', must meet at some point, as E. Let the forces R, R', be resolved into two rectangular components, respectively parallel and perpendicular to AD. The components perpendicular to AD are $R \cos a$, $R' \cos a$, which are equal to each other if the wedge is isosceles, and hence balance. The sum of those components of the resistances which are parallel to DA, must be equal to $2P$; that is, for equilibrium,

$$2P = 2R \sin a, \text{ or } P = R \sin a \quad (148); \text{ whence}$$

$$P : R :: \sin a : 1 :: BD : AB. \quad (149.)$$

Hence there is equilibrium with the wedge *when power is to weight as half the back is to the face of the wedge.*

224. Practical Applications. The wedge is generally used for purposes of cleavage, as in splitting timber. A cleft is made, into which the edge of the wedge is introduced, and the blows of a mallet are applied to drive it forward. As the theorems which have just been given apply to the case of equilibrium under pressures, while the wedge is usually driven by percussion, they are of very limited use, and hence of little practical use. Were it not for friction, which is not taken into account in their deduction, the wedge would lose most of its value. In fact, all that can be said is that the cleaving power of the wedge increases as its angle diminishes.

Most of our cutting instruments are applications of the wedge, as the chisel, saw, knife, razor, etc. The angle of the instrument depends upon the tenacity of the substance which is to be cut, as the diminution of the angle must practically be limited by the corresponding decrease of strength in the tool itself. Hence the harder the substance on which the tool is to be used, the less acute is the angle. Thus chisels for cutting wood have an angle of about 30°, for iron an angle of from 50° to 60°, and for brass from 80° to 90°.

VII. *Screw.*

225. Construction of the Screw. The *screw* is formed by winding a narrow inclined plane about a cylinder, as shown in Fig. 118. The spiral *ABCDEF* formed by the plane is called the *thread* of the screw, and works in a similar thread cut on the interior of a block called a *nut*, *PQ*, Fig. 119. On turning the screw around, it is forced to advance or recede in the nut, according to the direction of the motion. The power is generally applied at the extremity of a lever *AB*, Fig. 119, which is attached to the head of the screw. To find the relation between the power and the pressure exerted by the screw, we have simply to consider the case of a body *L*, Fig. 118, resting on an inclined plane. On turning the screw, *L* will be pushed up the plane with a certain force which is evidently equal to the force with which the screw will descend if *L* is fixed. This last supposition is evidently precisely the case of a screw with a fixed nut, for *L* may be supposed to embrace a considerable portion of the thread. The power *P* turning the screw is applied parallel to the base of the plane, hence from (142) $P : W :: H : B$. Now, considering only a single turn of the thread, this forms an inclined plane of which the base is the circumference of the cylinder on which the thread is formed, and the height is the distance between the threads. Hence, calling P' the power applied, W the pressure exerted by the screw, d the distance between the threads, and r the radius of the cylinder, $P' : W :: d : 2\pi r$, (149). Proportion (149) supposes the power to be applied at surface of the cylinder. If the lever *CN* (Fig. 120) be attached, a force *P* acting at *N* will produce a pressure P' at *M*, which may be determined from the law of moments. If *R* be the radius *CN*, r the radius *CM* of the cylinder, we have $PR = P'r$, whence $P' = P\frac{R}{r}$. Substituting this value of P' in (149) we have $P\frac{R}{r} : W :: d : 2\pi r$ or $P : W :: d : 2\pi R$ (150). That is, with the screw there is equilibrium when *the power applied to the lever is to the pressure produced by the screw as the distance between the threads of the screw is to the circumference described by that portion of the lever arm to which the power is applied.*

From (150) we obtain the equation $W = P\frac{2\pi R}{d}$ (151). The effect would evidently be the same if the screw were fixed and the nut movable, so that the proposition is general.

The preceding theorem is incorrect in practice, owing to the enormous friction produced by the threads.

226. Definitions. It will be seen that at each turn of the screw it moves over a space equal to the distance between the threads. This distance is known as the *pitch* of the screw. It is generally determined by noticing the number of ridges occurring in an inch of the length of the screw. The angle made with a horizontal line by a tangent to the thread of the screw is the *angle of the screw*. A screw generated by the revolution of a single band around a cylinder, as already described, is said to be *single-threaded*. To increase the strength of the instrument, it is not uncommon to form the screw by cutting two parallel threads side by side, making a *double-threaded* screw. This increase in the number of threads does not alter the pitch, which is dependent only on the angle of the screw and the size of the cylinder upon which it is cut. Two forms of screw-thread are used in practice, the *square thread*, Fig. 121, and the *V thread*, Fig. 122. The square thread is the most powerful because the reaction of its surface RR, Fig. 121, is parallel to the elements of the cylinder of the screw, so that the total force applied acts to overcome the resistance, while with the V thread these reactions are oblique to the elements of the cylinder, and hence a portion of the applied force is uselessly exerted in producing a pressure tending to burst the nut. The V thread is, however, stronger, because there is less material cut away in forming it. A deep screw is evidently less strong than a shallow one, but it is more durable, as the greater amount of bearing surface prevents it from wearing away so easily.

Screws are used in practice whenever it is necessary to exert a great pressure though a small space, as in pressing books, extracting oil from seeds, raising buildings, and the like. Boring instruments as augers, gimlets, and corkscrews, are additional examples.

227. Differential Screw. Equation (151) shows that the mechanical advantage of a screw may be increased in two ways; (1) by increasing the length of the lever arm, and (2) by diminishing the distance between the threads. The first of these methods greatly increases the size of the machine, while the second is practically limited, owing to the weakness of extremely fine threads. The difficulties are avoided in the machine known as the *differential screw*, or *Hunter's screw*, one of the forms of which is shown in Fig. 123.[1] Two screws AB, CD, are cut upon a single cylinder, the pitch of AB being greater than that of CD. AB works in a nut EF, and CP enters a second nut, which is capable of a vertical motion between the guides $DEFG$, and is prevented from turning around with the screw. To understand its working, suppose the pitch of AB to be $\frac{3}{5}$ of an inch, and that if CD to be $\frac{2}{5}$ of an inch. On turning AB once around, it advances through a space of $\frac{3}{5}$ of an inch. The smaller screw CD at the same time turns in the nut GH, and by its motion as it enters the nut draws this towards EF by an amount of $\frac{2}{5}$ of an inch. Hence by a single turn GH is moved away from EF $\frac{3}{5}$ of an inch by the motion of AB, while it is moved towards EF $\frac{2}{5}$ of an inch at each turn, and hence the same effect is produced as if there were a single screw with a pitch of $\frac{1}{5}$ of an inch, the difference between the pitch of AB and that of CD. The differential screw is therefore in equilibrium when *power is to pressure produced as the differences of the pitches of the two parts of the screw is to the circumference described by the power.*

Since the effect is dependent merely on the *difference* of the pitch and not on its absolute value, the ratio of pressure to power may evidently be in-

[1] From the inventor, John Hunter, the celebrated surgeon.

creased to any desired extent without diminishing the strength of the apparatus. Suppose, for example, the pitch of AB to be $\frac{1}{20}$ in., and that of CD $\frac{1}{30}$ in., and let the end of the lever-arm describe a circumference of 10 inches. A power of 10 lbs. applied at the extremity of the lever will cause a pressure of $(10 \times 50) \div (\frac{1}{20} - \frac{1}{30}) = 30,000$ lbs.

228. Endless Screw. Fig. 123 represents what is known as an *endless screw.* It consists of a wheel AB, called a *worm-wheel*, furnished with oblique teeth which engage in the thread of the screw CD, which is known as the *worm.* If the screw is single-threaded, the wheel advances by a single tooth for each revolution of CD. If there are two, three or more threads, a corresponding number of teeth pass at each revolution. The reduction of velocity taking place in the transfer of motion from the screw to the wheel, renders this contrivance useful for registering the number of revolutions of an axis. Thus if the wheel AB has 50 teeth, it will make but 1 revolution, while the screw makes 50. If, now, the axis AB carry a pinion acting in its turn on a cogged-wheel, the velocity of the revolution may be reduced to any desired extent. For instance, if the pinion has 5 teeth and the wheel 100, the screw will revolve 1000 times while the cogged-wheel revolves once. The number of revolutions of the screw-axis is read by means of an index attached to the axis of the last wheel.

When any considerable force is to be transmitted by the endless screw, the screw is made to drive the wheel, because the friction would prevent the wheel from driving the screw, in addition to the fact that such a combination would involve a mechanical disadvantage. In light machinery, however, where there is little friction, the wheel may be made the driver. This combination is frequently used to regulate the velocity of a train of wheels by connecting the axis of the screw with a fan-wheel, the motion of whose wings is so much resisted by the air as to keep the velocity within suitable limits.

229. Micrometer Screw. By means of the screw we are enabled to measure small linear distances with great accuracy. When the screw is turned, the space passed over by its extremity is known immediately from the number of revolutions, provided that we know the pitch, and if a graduated circle is attached to the screw-head a minute fraction of a turn can easily be measured. Suppose, for example, that a circle divided at its circumference into 100 parts be thus attached to a nicely-made screw having a pitch of 1 millimetre. If the screw makes one complete revolution, it will advance 1 millimetre, hence for $\frac{1}{100}$ of a revolution, the length of one of the divisions of the graduated circle, it will advance $\frac{1}{100}$ mm., which distance therefore corresponds to one of the scale-divisions. With a first-class micrometer screw we can measure a length within $\frac{1}{1000}$ of a millimetre.

230. Sheet Metal Gauge. A practical application of this principle is shown in the gauge used for measuring the thickness of sheets of metals, paper, etc., Fig. 124. "The piece in the form of the letter U has a projecting hub. a, on one end. Through the two ends are tapped holes, in one of which is an adjusting screw, B, and in the other the gauge-screw, C. Attached to the screw C is a thimble, D, which fits over the exterior of the hub, a. The end of this thimble is beveled, and the beveled edge graduated into 25 parts, and figured 0, 5, 10, 15, 20. A line of graduations 40 to the inch is also made upon the outside of the hub a, the line of these divisions running parallel with the centre of the screw C, while the graduations on the thimble are circular. The pitch of the screw C being 40 to the inch, one revolution of the thimble opens the gauge $\frac{1}{40}$ or $\frac{25}{1000}$ of an inch. The divisions on the thimble are then read off for any additional part of a revolution of the thimble, and the number of such divisions is added to the turn or turns already made by the thimble, allowing $\frac{25}{1000}$ for each graduation on the hub a. For example, suppose the thimble to have made 4 revolutions and one-fifth. It will then be noticed that the beveled edge has passed four of the graduations on the hub a, and opposite the line of graduation will be found on the thimble the line marked 5. Add this number to the amount of the four graduations which is $\frac{100}{1000}$, and equals $\frac{105}{1000}$, which is the measurement shown by the gauge.[1]" To show the method of using the instrument, suppose we have to determine the thickness of a sheet of metal EF. We first turn the screw till the ends B and C are in contact. The reading on the scale should then be 0. Then turning the screw till the sheet will pass between B and C, we place it as shown in the figure, and again revolve the screw until the end just touches the metal, taking the reading of the scale as before. The second reading evidently shows the number of revolutions made by the screw in passing over a linear distance equal to the thickness of the sheet of metal, and this multiplied by the pitch of the screw is the thickness sought. If the material of the sheet is very soft, it is placed between two plates of glass, the reading being taken when the metal is in place and after it has been removed. The difference of the readings gives the thickness sought.

231. Dividing Engine. The micrometer screw is also applied in the construction of the engines used for dividing linear scales. The principle of the instrument is as follows. A diamond, or other graver, is attached to an arm, which is connected with the nut of a carefully-made screw, so as to be carried forward when this is revolved. Directly under the graver, and parallel to the line of its motion, is placed the bar to be divided. The graduated head of the screw is turned through a certain number of divisions corresponding to the spaces to be marked off upon the bar. A movement is then given to the graver, cutting a line upon the bar, the screw is again revolved, and thus the process is continued. The best dividing engines have devices for rendering the whole action automatic, and also for making every fifth and tenth mark on the scale larger than the intermediate ones. Oftentimes the graver is stationary, and the bar to be graduated is moved by the screw. A similar device is used in the simplest method of dividing circles. The plate to be graduated is attached concentrically to a heavy circular wheel, to which a tangential micrometer screw of known pitch (*tangent-screw*) is fitted. The rotation of the tangent-screw through a single turn corresponds to the rotation of the wheel and the attached plate through a certain fraction of a degree. At each turn of the screw a mark is made

[1] Catalogue of Darling, Brown and Sharpe.

upon the plate by the graver, and the process is continued until the whole circumference has been traversed and completely graduated. Additional examples of the use of the micrometer screw occur in *reading-microscopes*, the *spider-line micrometer*, etc.

232. Combinations of the Mechanical Powers. The mechanical powers may be combined in various ways. The mechanical advantage gained by any particular combination may be determined by estimating the effect of its component parts separately. For example, Fig. 125 represents a form of machine sometimes used for raising ships from the water. The vessel is drawn up an inclined plane by means of the wheel and axle (*capstan*) HG, acting upon the system of movable pulleys EF. If we suppose the radii GH, GI, to be 2 metres and one-fourth metre respectively, and the slope of AC to be 1 metre in 10, the power exerted at K, on applying at H a force of 1000 kgrs., will be 32,000 kgrs., which would be sufficient to draw a vessel weighing 320,000 kgrs. up the plane AC, supposing no power to be lost by friction.

VIII. *Application of the Principle of Virtual Velocities to the Mechanical Powers.*

233. General Principles. In the use of the various mechanical powers there is no absolute gain of force. For it will be recollected that the force of motion of any body is equal to its mass multiplied by its velocity, and it is found that if the power and weight are in equilibrium, and a motion is impressed upon the machine in the direction of either of those forces, the power multiplied by its velocity equals the weight multiplied by its velocity; or, calling P the power, W the weight, V, V' their respective velocities. $PV = WV'$ (152), whence $P : W :: V' : V$. (153.) Hence, the velocity of the weight when compared with that of the power, is diminished in the same ratio as the weight balanced by that power is increased by the use of the machine, or as it is generally expressed, *what is gained in power is lost in time.*

The demonstration of this principle is simply an application of the general theorem of *Virtual Velocities*. (§ 141, p. 73.) We proceed to consider the various mechanical powers in order. Our method of proof will be to show that assuming the principle of virtual velocities, the law of equilibrium as deduced for each power follows.

234. Application to Lever. Let us take the case of a bent lever, AFB, Fig. 126, P being the power, and W the weight balancing it. Suppose now that a slight displacement be given to the machine, so that A takes the position A', and B moves to B'. The power now acts along $A'P'$, the weight along $B'W'$. Draw $A'm$ perpendicular to AP, and $B'n$ perpendicular to the prolongation of WB. Am and $-Bn$ are the virtual velocities of P and W, Am being positive, as it is laid off in the direction of the action of P, and Bn negative, because it is laid off opposite to the direction of the action of W. The arcs AA', BB', being very small, may be considered as straight lines perpendicular to AF and BF, respectively. By the principle of virtual velocities,

$$P \times Am - W \times Bn = 0. \quad (154.)$$

Now $Am = AA' \cos mAA'$, also arc $AA' = AF \times$ angle AFA', and $\cos mAA' = \cos (PAF - 90°) = \sin PAF$. Hence $Am = AF \times$ angle $AFA' \times \sin PAF$.

In like manner, $Bn = BB' \cos nBB'$. But $BB' = BF \times$ angle BFB', and $\cos nBB' = \cos(WBF - 90°) = \sin WBF$. Hence $Bn = BF \times$ angle $BFB' \times \sin WBF$. Substituting these values in equation (154), $P \times AF \times$ angle $AFA' \times \sin PAF - W \times BF \times$ angle $BFB' \times \sin WBF = 0$. (155.)

But $AFA' = BFB'$, from the conditions of motion, whence $P \times AF \times \sin PAF - W \times BF \times \sin WBF = 0$ (156), or

$$P : W :: BF \sin WBF : AF \sin PAF \quad (157),$$

which is the condition of equilibrium, as deduced for the lever in proportion (101), p. 88, as $AF \sin PAF$, $BF \sin WBF$, are the arms of the forces P and W.

235. Application to Wheel and Axle. Let ACB, Fig. 127, represent a wheel and axle. The forces applied are P and W, acting vertically downwards, and the upward reaction $R = P + W$, at C. The machine being rigid, the forces may be considered as applied in the same plane. The virtual velocity of $R = 0$, as the wheel and axle revolve about C. Suppose a motion impressed upon the machine so that P moves to P', thus raising W to W'. The original point of contact of the rope P is A, and the wheel moves from A to A', and the original point of contact of the rope W is B, and the axle moves from B to B'. $PP' = AA'$ is the virtual velocity of P, and $-WW' = -BB'$, that of W. For equilibrium,

$$P \times AA' - W \times BB' = 0. \quad (158.)$$

But $AA' = AC \times$ angle ACA', and $BB' = BC \times$ angle BCB', or, calling R, r the radii of the wheel and axle, $AA' = R \times$ angle ACA', $BB' = r \times$ angle BCB'. Substituting these values in (158), $P \times R \times$ angle $ACA' - W \times r \times$ angle $BCB' = 0$, or, as the angles ACA' BCB', are equal, $P \times R - W \times r = 0$, or $P : W :: r : R$ (159), the condition of equilibrium demonstrated in § 193, p. 97.

236. Application to Pulley. With a fixed pulley and parallel cords, the truth of the principle is evident, as the power and weight move over equal distances.

The most general case of the movable pulley is that in which the cords are inclined, as represented in Fig. 128. Let C, P, W, be the original positions of the pulley, power and weight. On impressing a slight motion upon P, so that it moves to P', C rises to C', and W to W', so that $WW' = CC'$. PP' is the virtual velocity of P, $-WW'$ that of W. Draw Cm, Cn, circular arcs having their centres at M and H. When the displacement is very small these become straight lines perpendicular to Hn, Hm. In that case $Cm = Cn = CC' \cos HCC' = CC' \cos \frac{1}{2}HCM = CC' \cos a$, using the notation of § 208, p. 102. Since the shortening of the whole of the cord MCH, is equal to the lengthening of MP, $PP' = mC + nC = 2CC' \cos a$. For equilibrium, $P \times PP' - W \times CC' = 0$. (160.) Substituting the above value of PP',

$$P \times 2CC' \cos a = W \times CC', \text{ whence}$$

$W = 2P \cos a$ (161), which is identical with equation (127), p. 103.

237. Application to Inclined Plane. Let A, Fig. 129, be the original position of the point of application of F and W, and suppose it to be displaced along the line AB to the point B. Then Am is the virtual velocity of F, and $-Bn$ that of W. For equilibrium, $F \times AM - W \times Bn = 0$. (162.) But $Am = AB \cos B$, and $Bn = AB \sin a$; hence, substituting these values in (162),

$$F \times AB \cos B - W \times AB \sin a = 0, \text{ or}$$

$$F : W :: \sin a : \cos B \quad (163), \text{ as demonstrated in § 216, p. 106.}$$

238. Application to Wedge. Calling $2P$ the power, and R, R', the equal reactions, we have in the case of a small displacement of the wedge from the position ABC to $A'B'C'$, Fig. 130, *virtual velocity of* $P = mm' = AA'$, *virtual velocity of* $R = -nn' = -Ap$. If there is equilibrium, $P \times AA' - R \times AP = 0$ (164), or

$$P : R :: Ap : AA'.$$ But

$Ap : AA' :: BN : BA$, whence $P : R :: BN : BA$ (165), which is identical with the conditions of equilibrium demonstrated in § 223, p. 110.

239. Application to Screw. The application of the theorem to the screw is obvious, since the power must move through the circumference whose radius is the lever-arm, while the weight moves over a distance equal to the pitch of the screw. For any small displacement the virtual velocities would be in this ratio, whence

$P : W :: d : 2\pi R$ (166), as already demonstrated.

240. Velocity-Ratio. The ratio of the virtual velocities of the power and weight, $\frac{V}{V'}$, is called the *velocity-ratio* of those forces. From what has already been demonstrated, will be seen that in all cases *the mechanical advantage gained by the use of any machine is expressed by the velocity-ratio of the power and weight.* This theorem gives a simple method of determining the theoretical efficiency of any combination of the mechanical powers.

References.

Treatise on Mechanics, by Capt. Henry Kater and the Rev. Dionysius Lardner; Chap. xxi., *On Balances and Pendulums.*

Reports of U. S. Commissioners to the Paris Universal Exposition, 1867. Vol. III. *Machinery and Processes of the Industrial Arts, and Apparatus of the Exact Sciences*, by F. A. P. Barnard; p. 484. *Balances.*

Mount Washington in Winter, Chap. iv.; (Boston, 1871.) *Description of Mt. Washington Railroad.*

Die Righi-Eisenbahn mit Zahnradbetrieb, beschrieben von Prof. J. H. Kronauer.

Etude sur les Chemins de Fer de Montagnes avec Rail á Crémaillère, par M. A. Mallet (Paris, 1872).

Article on Mt. Cenis Railroad, Journal of the Franklin Institute, Vol. L., p. 289.

Mountain Railways; Van Nostrand's Eclectic Engineering Journal, Vol. VII, p. 407. (From *Engineering.*)

Centre Rail System, Do., Vol. VII, p. 394. *Account of Cantagallo Railroad.* (From *Engineering.*)

Account of Slide of Alpnach; Works of Professor John Playfair, Vol. I. (Edinburgh, 1822.)

For various forms of Dividing Engines consult references given on p. 24.

Also consult *Reports of U. S. Commrs. to Paris Universal Exposition*, Vol. III, Chap. xviii., *On Metrology and Mechanical Calculation*, for various forms of *comparators*, *spherometers*, etc.

See also the same volume, p. 11, for a description of *Whitworth's Micrometric Apparatus.*

CHAPTER XI.

Uniform and Uniformly Variable Motion.

241. Definitions. *Dynamics* treats "of the relations between the motions of bodies and the forces acting amongst them."[1] It differs from *Kinematics* in that the latter subject considers only the relations of the motions to each other, independently of their causes.

Motion is of two kinds: (1) *motion of translation*, in which the moving body simply passes from point to point, the paths of all the particles composing it lying in parallel lines, and (2) *motion of rotation*, in which the body under consideration revolves about an axis. These two varieties of motion frequently exist in combination. A motion of translation may be *rectilinear*, in which case the path of the moving body is a straight line; or *curvilinear*, when the path is any curve.

All motion is either *uniform* or *variable*. In uniform motion equal spaces are described in equal times; in variable motion unequal spaces are described in equal times.

242. Uniform Motion: Formulæ. Since a body having a uniform motion describes equal spaces in equal times, if in a unit of time it describes a space V, in 2 units it describes a space $2V$, and in t units a space tV. Hence, denoting by S the whole space passed over by the body in t units of time, with a constant velocity V, we have,

$$S = Vt \quad (167), \text{ whence } t = \frac{S}{V} \quad (168), \text{ and } V = \frac{S}{t} \quad (169),$$

the formulæ for uniform motion.

243. Causes producing Uniform Motion. There are two cases in which uniform motion may occur. (1.) If a body were projected with a definite velocity, in the absence of all disturbing forces, it would proceed indefinitely with a uniform motion, by virtue of its inertia. Practically, these conditions can never be exactly fulfilled, but as they are approximately realized, uniformity of motion is also approached. (2.) When the resultant of all the forces acting upon the moving body is zero, that is, when the magnitude of the impelling forces equals the magnitude of the resisting forces, the motion is also uniform. For an example, let us take the case of a train of cars. Here the train is moved onward by the impelling force of the locomotive, while the friction of the wheels is a constant retarding force. Suppose the train to be in motion. Were there no retarding force, its inertia

[1] Rankine, *Applied Mechanics*, p. 475.

would carry it onward with undiminished velocity. But the friction on the rails, and other resistances, destroy a portion of the motion in each unit of time, and if acting alone would finally consume the whole motion, when the train would come to rest. The locomotive, however, gives a constant impelling force, and if this is made just equal to the resistance, the loss of motion by the latter is compensated by the gain from the former, and the velocity of the train is unaltered.

A body thus moving uniformly under the action of balanced forces is said to be in *dynamical equilibrium.*

244. Uniformly Accelerated and Uniformly Retarded Motion, and their Causes. When the spaces traversed in successive equal times increase, the motion is said to be *accelerated*; when they decrease it is said to be *retarded.* If the increments or decrements of motion are equal in equal times, we have *uniformly accelerated* or *uniformly retarded motion.* When a moving body is acted upon by any force exterior to itself, its motion is caused to vary. If this force acts in the direction in which the body is moving, the velocity will be increased, thus producing accelerated motion; if the force is opposed to the motion of the body, the velocity will be decreased, producing retarded motion. If the force acting upon the body is constant, equal increments or decrements of motion will be generated or destroyed in equal times, causing uniformly accelerated or uniformly retarded motion.

245. Definition of Velocity. In uniformly accelerated or retarded motion the velocity of the moving body is constantly varying. The velocity at any instant may therefore be defined as the space over which the body would move in a unit of time were the accelerating or retarding force to cease its action.

In uniformly accelerated motion the increment of velocity in a unit of time is called the *acceleration.* In uniformly retarded motion the decrement of velocity in a unit of time is called the *retardation.* The unit of time generally adopted is the second.

246. Laws of Uniformly Accelerated Motion. Formulæ. I. *The velocity acquired by a body moving with a uniformly accelerated motion is proportional to the time.* For by the definition of this kind of motion, the increments of motion are equal in equal times. Therefore if a body under the action of any force acquires a velocity, a, in 1 second, in 2 seconds it will acquire twice that velocity, or $2a$, and in t seconds t times a. Hence calling a the acceleration, t the time, and v the velocity at the end of that time, we have $v = at$ (170), a formula giving the relation between the acceleration, time and velocity, and showing that the velocity varies directly as the time.

II. *The successive spaces traversed by a body moving with a uniformly accelerated motion are proportional to the squares of the time occupied in passing over them.* The space traversed by a body during an extremely small interval of time may be regarded as moved over with a uniform motion. Let t be the time which the body occupies in passing over a space s, and let that time be divided into n equal parts, n being a very large number. Each one of these parts will equal $\frac{t}{n}$. Calling a the acceleration, the velocities after successive intervals of time, $\frac{t}{n}$, $2\frac{t}{n}$, $3\frac{t}{n}$, $\cdots\cdots$ $n\frac{t}{n}$ will be, according to equation (170), $a\frac{t}{n}$, $2a\frac{t}{n}$, $3a\frac{t}{n}$, $\cdots\cdots$ $na\frac{t}{n}$, and supposing the velocity to be uniform during each of these intervals of time, the corresponding spaces passed over will be $a\frac{t^2}{n^2}$, $2a\frac{t^2}{n^2}$, $3a\frac{t^2}{n^2}$, $\cdots\cdots$ $na\frac{t^2}{n^2}$, by equation (167). The space s traversed in time t will evidently be the sum of these small intervals, therefore $s = a\frac{t^2}{n^2} + 2a\frac{t^2}{n^2} + 3a\frac{t^2}{n^2} + \cdots\cdots\, na\frac{t^2}{n^2}$, or $s = a\frac{t^2}{n^2}(1 + 2 + 3 + \cdots\cdots n)$. The quantity within the parentheses is the sum of the terms of an arithmetical progression, and $= \frac{(n+1)n}{2}$, hence $s = a\frac{t^2}{n^2}\left[\frac{(n+1)n}{2}\right]$, or $s = \frac{at^2}{2} + \frac{at^2}{2n}$. A consideration of the cause of uniformly accelerated motion will make it evident that only when the interval of time considered is infinitely small, that is, when n is infinitely great, does the supposition which has been made, that the spaces traversed during the interval $\frac{t}{n}$ are uniform, apply to the kind of motion with which we are now concerned. Hence, that the equation just deduced may be applicable to uniformly accelerated motion, n must be made $= \infty$. Inserting this value therein, the formula becomes, $s = \frac{at^2}{2} + \frac{at^2}{\infty} = \frac{at^2}{2} + 0$, whence $s = \frac{1}{2}at^2$ (171), a formula showing the relation between the space traversed, the time occupied in passing over it, and the acceleration, and from which it follows that s varies as t^2, or the spaces are proportional to the squares of the times.

III. *In uniformly accelerated motion the velocities are proportional to the square roots of the spaces passed over in generating them.* For from (170) $v = at$, and $t = \frac{v}{a}$, and from (171) $s = \frac{1}{2}at^2$, and $t = \sqrt{\frac{2s}{a}}$. Equating these two values of t, we have $\frac{v}{a} = \sqrt{\frac{2s}{a}}$, or solving relatively to v, $v = \sqrt{2as}$ (172.), in which v varies as the square root of s.

IV. *In uniformly accelerated motion the spaces passed over in successive equal intervals of time are proportional to the odd numbers,* 1, 3, 5, 7, 9, etc. This follows from formula (171), $s = \frac{1}{2}at^2$. For in intervals of time 1, 2, 3, 4, 5, the spaces traversed are $\frac{1}{2}a$, $\frac{4}{2}a$, $\frac{9}{2}a$, $\frac{16}{2}a$, $\frac{25}{2}a$. The spaces travelled over in each successive interval will be found by subtracting these spaces, $\frac{1}{2}a$ from $\frac{4}{2}a$, $\frac{4}{2}a$ from $\frac{9}{2}a$, $\frac{9}{2}a$ from $\frac{16}{2}a$, etc., and are respectively equal to $\frac{1}{2}a$, $\frac{3}{2}a$, $\frac{5}{2}a$, $\frac{7}{2}a$, $\frac{9}{2}a$, quantities which are to each other as 1, 3, 5, 7, 9, and the same reasoning can evidently be applied to a greater number of successive intervals.

The same fact may also be shown to follow from a consideration of the first and second laws of motion in connection with formula (171). The space passed over in a unit of time is $\frac{1}{2}a$, and the velocity at the end of that time is a. By virtue of its inertia alone, in a second unit of time the body would pass over a space a, while the additional space traversed, owing to the continued action of the accelerating force is $\frac{1}{2}a$. The total distance travelled over during the second unit of time is therefore $\frac{3}{2}a$. The velocity at the end of this second interval is $2a$. During a third interval the body will pass over a space $2a$, because of its inertia, and $\frac{1}{2}a$, because of the continued action of the force, in all $\frac{5}{2}a$. The same reasoning could be applied to a greater number of equal intervals, showing that in the fourth the space is $\frac{7}{2}a$, in the fifth $\frac{9}{2}a$, and so on. In intervals 1, 2, 3, etc., the spaces traversed therefore vary as the numbers 1, 3, 5, 7, 9.

247. Body projected in direction of Accelerating Force. If the body has an initial velocity, V, the velocity at the end of time t, will be $v = V + at$. (173.) For by the second law of motion the accelerating force will produce the same effect as if the body were at rest, hence the total velocity will be that due to the inertia of the body plus the velocity generated by the accelerating force.

The space described in time t, will be $s = Vt + \frac{1}{2}at^2$. (174.) For the space traversed by virtue of the initial velocity will be Vt, while that described because of the acceleration is $\frac{1}{2}at^2$. The total space is therefore equal to the sum of these two.

248. Formulæ for Uniformly Retarded Motion. In uniformly retarded motion the constant force acts in opposition to the original motion of the body, thereby tending to bring it to a state of rest. The amount of motion destroyed in a unit of time by a given force acting in opposition to a preëxisting motion, is equal to that which the same force would generate in a body starting from a state of rest under its influence. From this it follows that at the end of a time t, the velocity of a body moving with a uniformly retarded motion is $V - at$ (175), a being the retardation, and V the initial velocity. For as a is the quantity of motion destroyed in 1 unit of time, the velocity, at, must be destroyed in t units.

The space traversed in time t, will be $s = Vt - \frac{1}{2}at^2$ (176), which is the distance Vt, over which the body would move by virtue of its inertial velocity V, *minus* the space $\frac{1}{2}at^2$, through which it would move in an opposite direction, were it solicited from a state of rest by the retarding force. This is a direct consequence of the second law of motion.

249. Laws of Uniformly Retarded Motion. I. From these formulæ it follows that *the space described in extinguishing a given velocity is equal to that described in the generation of the same velocity under the action of the same force.* For the time required to destroy a given motion equals the time required to generate that motion. Let V be the given initial velocity, which is supposed to be extinguished in time t. Then $V = at$. By the formula (176) the space $s = Vt - \frac{1}{2}at^2$. Substituting in this the above value of V, $s = at^2 - \frac{1}{2}at^2 = \frac{1}{2}at^2$, which is also the space described in time t, by a body starting from rest, and moving with a uniformly accelerated motion.

II. *The velocity at any particular point of the space described is the same whether the body starts from rest and acquires a velocity v, or starts with a velocity v and is gradually brought to rest.* Let AB, Fig. 131, be the path described with a uniformly accelerated motion, and BA that described with a uniformly retarded motion. When moving from B to A the velocity possessed by the body on arriving at any point C, is just sufficient to carry it up to A. From the preceding proposition it follows that this velocity is exactly equal to that which it would acquire by passing over AC, under the action of the force causing the acceleration.

250. Case of Falling Bodies. A body falling vertically through the air moves with a uniformly accelerated motion, as it is acted upon by a constant accelerating force, its weight, or gravity, pulling it towards the earth. A body thrown vertically upward moves with a uniformly retarded motion, the weight in this case acting in opposition to the projecting force. The acceleration produced by gravity is 9.8 m., and is usually denoted by the letter g. Substituting this letter in formulæ (170), (171), (172), (173), (174), (175), (176), and for s writing h the *height*, we obtain formulæ for the solution of all numerical questions relative to bodies falling freely under the influence of gravity, as follows: —

$v = gt$ (177), $h = \frac{1}{2}gt^2$ (178), $v = \sqrt{2gh}$. (179.)
$v = V + gt$ (180), $h = Vt + \frac{1}{2}gt^2$. (181.) } Formulæ for uniformly accelerated motion.

$v = V - gt$ (182), $h = Vt - \frac{1}{2}gt^2$. (183.) } Formulæ for uniformly retarded motion.

251. Application of Formulæ. The various experimental verifications of the laws of motion deduced in the present chapter will be found in the discussion of falling bodies, the motion caused by gravity being the variety of accelerated motion most easy to deal with. The application of the preceding formulæ will be made clear by the consideration of the following numerical examples.

1. Suppose that a body falls freely for 5 seconds, and we wish to determine the velocity at the end of that time, and also the total space traversed. Substituting in (177), $g = 9.8$, $t = 5$, we have, $v = 9.8 \times 5 = 49.0$ m. Also from (178), $h = \frac{1}{2} \times 9.8 \times (5)^2 = 122.5$ m. Or h could be found directly from v, as $v = \sqrt{2gh}$, whence $49 = \sqrt{2 \times 9.8 \times h}$. Solving this relatively to h, the same value as before, $h = 122.5$ m. is obtained.

2. Suppose that the body were projected vertically downward with a velocity of 10 m., and that we wish to know its velocity after 5 seconds, and the space traversed in that time. From (180) $v = V + gt = 10 + 9.8 \times 5 = 59$ m.; and from (181.), $s = Vt + \frac{1}{2}gt^2 = 10 \times 5 + \frac{1}{2} \times 9.8 \times (5)^2 = 172.5$ m.

3. A body is projected vertically upward with a velocity of 10 m., and it is required to find the time it will be in the air, and the height to which it will ascend. The time is found from (177), since the time occupied in destroying a velocity v equals that necessary to generate it. Substituting for v and g their values, we have $100 = 9.8\ t$, or $t = \frac{100}{9.8} = 10.2$ sec., which is the time of ascent, and as the times of ascent and descent are equal, twice this time, or 20.4 is the total time that the body remains in the air. The height is found either from (179) or (183). For $v = \sqrt{2gh}$, or $100 = \sqrt{2 \times 9.8 \times h}$, whence $h = 510.2$ m. Or from (183), $h = Vt - \frac{1}{2}gt^2$, $h = 100 \times 10.2 - \frac{1}{2} \times 9.8 \times (10.2)^2 = 510.2$ m.

252. Graphical Representation of Motion. Curve of Spaces. It is often convenient to study the laws of motion by the use of the Graphical Method.

Any rectilinear movement of a particle is defined when we know its distance from some fixed point of reference, at each successive unit of time. Hence we draw two rectangular axes, OX, OY, Fig. 132, representing successive units of time, by equal distances measured on OX (abscissas), and denote by vertical lines (ordinates) the distance of the moving body from the point of reference O at the end of each instant. By joining the extremities of these ordinates we may construct a curve representing the motion of the body. For example, let us suppose that we wish to exhibit graphically the law of motion of a particle vibrating to and fro about a point D, Fig. 133. Let the particle occupy the positions B''', B'', B', B, D, C, C', C'', C''', at the expiration of successive units of time, then returning through the points C''', C'', C', C, D, and so on. We assume the middle point, D, as the point to which the motion is to be referred, and lay off equal lengths, OB, BB', $B'B''$, $B''B'''$, etc., along the axis OX (Fig. 132), representing successive units of time. Those spaces at the left of O are negative, and denote intervals of time before the particle reaches D; those on the right are positive, and denote times after the particle has passed D.

Now from each of these points, B''', B'', B', etc., we erect ordinates of length proportional to the distances DB''', DB'', DB', etc. (Fig. 133), representing positions at the left of D by negative ordinates, and those at the right by positive ordinates. Drawing a curve through the extremities of these ordinates, we obtain a graphical representation of the motion of the particle. From this curve it will be seen that, for example, 3 units of time before reaching D the particle is distant from it by a space represented by $B''S$. The distance from D, two units after passing it, is represented by $C'M$, 5 units by $C^{iv}N$, and so on, the ordinates first increasing as the particle recedes from D, then decreasing as it reapproaches D, and again becoming negative when D is passed, and the particle moves from D towards B''. Having constructed the curve we can readily obtain intermediate values. Thus if we wish to find the distance of the particle from D at the end of $3\frac{1}{2}$ units of time, we have simply to lay off $OK = 3\frac{1}{2}$, and ascertain the length of the corresponding ordinate KL, which represents the space required.

The curve constructed as explained, is known as the *Curve of Spaces.*

253. Determination of Velocity from Curve. From the curve of spaces it is easy to find the velocity possessed by the particle at any given distance from D in the following manner. In the case of variable motion the velocity at any time may be considered as uniform, while the particle is passing over a very small distance. Let VW, Fig. 134, be a portion of the curve of spaces. Suppose B', M to be periods of time taken so near together that the motion of the particle is sensibly uniform during the interval. The space traversed in this time is represented by OK, the time occupied in describing it by $B'M = NK$. Hence the velocity sought, $v = \frac{\text{space}}{\text{time}} = \frac{OK}{NK} = \frac{TL}{NL}$. But $NL = BB'$ represents a unit of time. Hence $v = TL$, that is, TL is the space which the particle would describe in a unit of time were it to continue moving with the velocity possessed by it at the given instant.

Since $\frac{OK}{NK} = \text{tang } TNL$, *the velocity at any particular time is the tangent of the angle made with OX by the geometrical tangent to the curve at the point corresponding to that time.*

254. Curve of Velocities. If we find the velocity corresponding to each successive unit of time, we may construct a *curve of velocities*, representing times by abscissas, and the corresponding velocities by ordinates, as in Fig. 135. Motions from left to right are positive, those from right to left negative.

If we have the curve of velocities given to construct the curve of spaces, this is easily done, for the change of distance of the particle from the point of reference during an extremely small portion of time is $CM \times MN = \textit{Area } CSMN$. The same method will give the space traversed in any other element of time. Hence *the whole space traversed by the particle in a given time is equal to the area of the corresponding portion of the curve of velocities.*

From the curve of velocities we can ascertain the *acceleration* exactly as we have already ascertained the velocity from the curve of spaces, since the acceleration is simply the increase of velocity in a unit of time.

255. Application to Uniformly Accelerated Motion. As a simple application of the preceding principles, we proceed to apply them to the demonstration of the laws of uniformly accelerated motion, which we have already proved by analytical methods. As the velocity is proportional to the time we construct the curve of velocities by laying off successive equal distances, OA, AB, BC, etc., on OX. These represent times. From each of the points A, B, C, etc., we raise perpendiculars AA', BB', CC', etc., of lengths proportional to the corresponding times, that is, we make $AA' : BB' :: OA : OB$, and so with all the rest. The line OE' is the curve of velocities, which, in the case of uniformly accelerated motion, is reduced to a straight line. From what we have already said the space traversed in any given time equals the corresponding area of the curve of velocities. Hence the space described in an interval of time denoted by OE is equal to the area of the triangle $OEE' = OE \times \frac{1}{2}EE'$. If $OA =$ 1 second, in which case AA', the velocity acquired in that time, becomes the acceleration, and $OE = t$, we have $OE : OA :: EE' : AA'$, or $t : 1 :: EE' : a$, whence $EE' = at$. Hence area $OEE' = OE \times \frac{1}{2}EE' = t \times \frac{1}{2}at = \frac{1}{2}at^2$.

256. Comparison of Forces with Gravity. The acceleration which will be produced when any force acts upon a body at liberty to move is readily found from proportion (7), (p. 35.) Calling W the weight of the body, and F the force, expressed in kilogrammes, a the acceleration produced by F, and g the acceleration produced by gravity, $F : W :: a : g$, whence $a = g\frac{F}{W}$ (184), a formula giving the acceleration when the force is known.

For example, suppose a body weighing 20 kgrs. to be acted upon by a force of 10 kgrs., and it is required to find the acceleration. From (184) we have $a = 9.8 \times \frac{10}{20} = 4.9$ m.

257. Motion down an Inclined Plane. We now proceed to investigate the subject of the motion of a body rolling down an inclined plane.

Let A, Fig. 137, represent any body resting upon an inclined plane of which $NO = H$ is the height, and $MN = L$ the length. The body presses in a vertical direction with a force equal to W its weight. Let this force be resolved into two components P and F, one of which, P, is perpendicular to the plane, and can have no tendency to cause the body to descend, while the component F, on the other hand, is parallel to the plane, and therefore exerts its whole influence to generate a motion down MN. By similarity of triangles $F : W :: NO : MN :: H : L$ whence $F = W\frac{H}{L}$ (185), a constant force. The body must therefore descend the plane with a uniformly accelerated motion. We wish to find the velocity and space in terms of the time and acceleration.

From (184) we have $a = g\frac{F}{W}$, and as $F = W\frac{H}{L}$, $a = g\frac{H}{L}$.

Substituting this value of a in the general formulæ (170), (171), (172), we have

$$v = g\frac{H}{L}t \ (186),\ s = \tfrac{1}{2}g\frac{H}{L}t^2 \ (187),\ v = \sqrt{2g\frac{H}{L}s} \ (188),$$

as formulæ for motion down inclined planes.

Since $\frac{H}{L} = \sin a$, the above equations may be written,

$$v = gt \sin a \ (189),\ s = \tfrac{1}{2}gt^2 \sin a \ (190),\ v = \sqrt{2gS \sin a} \ (191).$$

258. Velocity Acquired in Rolling down Inclined Plane. If we make $S = L$, equation (188) becomes $v = \sqrt{2gH}$, (192), which is the velocity acquired by a body rolling down the whole length L, of the plane, and which is independent of L, as that quantity does not enter into the value of v. But this is also the velocity acquired by a body falling vertically through the height H of the plane. Therefore *the velocity acquired in descending any inclined plane is independent of its length, and equal to the velocity acquired by falling through its vertical height.*

This proposition is also true for any portion of the plane, as BC, Fig. 138. For the velocity acquired in rolling over AB, is equal to that gained in falling through Am, and that acquired in rolling over AC to that gained in falling through An. Hence the velocity gained in passing over $AC—AB = BC$, is equal to that gained in falling through $An—Am = mn$.

259. Velocity acquired in descending a Series of Inclined Planes. *The velocity acquired by descending a series of inclined planes is equal to that acquired by falling through their perpendicular height.* Let AB, BC, CD, Fig. 139, be a series of inclined planes over which the body rolls from A to D. From what has already been shown, it is clear that the velocity at B is equal to that which would be acquired in falling through AM; in like manner the velocity acquired in rolling over BC is equal to that generated in falling through mn, and that acquired in rolling over CD to that gained in falling through nE. Hence, supposing that no velocity is lost by the change in direction of the motion at B and C, the whole velocity generated in passing over $AB + BC + CD$ is equal to that generated in falling through the vertical distance $Am + mn + nE = AE$, which is the total height of the series of planes.

260. Body descending a Curve. If the number of planes becomes infinite, the broken line $ABCD$ becomes a curve, from which it follows that *the velocity acquired in descending any portion of a curve is equal to that acquired in falling through its vertical height.*

261. General Proposition. Therefore in general, a *body by descending from a given height to a given horizontal line will acquire the same velocity whether the descent is made perpendicularly or* *obliquely, over an inclined plane, a series of inclined planes, or a curved surface.*

262. Investigation of Loss of Velocity. The preceding demonstrations suppose that there is no change of velocity produced when the body passes from one of the planes to the following one. It is important to ascertain the conditions under which this supposition is correct. Suppose the body to pass from AB to BC, Fig. 140, and denote the angle ABM by θ, and the velocity on reaching B by v. Let this velocity v be resolved into two components, one parallel to BC, the other perpendicular to it, and equal to $v \cos \theta$ and $v \sin \theta$, respectively. The perpendicular component $v \sin \theta$ will totally be consumed in producing a pressure upon BC, while the motion over BC will be wholly due to the parallel component $v \cos \theta$. Hence the loss of velocity $= v - v \cos \theta = v(1 - \cos \theta)$. That this quantity may equal 0, $\cos \theta$ must $= 1$, in which case $\theta = 0°$. This can only occur when the broken line ABC becomes a *curve*, of which AB and BC are elements. There is always, therefore, some loss of velocity except in the case of a body rolling down a curved surface.

263. Time of Descending an Inclined Plane. *The time occupied in passing over any inclined plane is to the time occupied in falling through its vertical height, as the length is to the height.*

Let t_l, t_h, be the times occupied in falling over the length L and the height H, respectively, and v the velocity acquired thereby. Then from (177) $v = gt_h$, and from (186) $v = g\frac{H}{L}t_l$. Equating these two quantities, $gt_h = g\frac{H}{L}t_l$, from which it follows that

$$t_l : t_h :: L : H. \quad (193.)$$

264. Time of Descent down Chords of Circle.

If a chord be drawn from either extremity of the vertical diameter of a circle, the velocity acquired in falling over it is proportional to the length of the chord, and the time of falling over any such chord is independent of its length and equal to the time of falling through the vertical diameter.

Let DC, Fig. 141, be any chord drawn from the extremity of the vertical diameter AC. Draw DB perpendicular to AC, and join AD. Denote AC by D, BC by H, and DC by L. The velocity acquired in passing over DC is $v = \sqrt{2gH}$. But as ADC is a right-angled triangle, $BC = \frac{DC^2}{AC}$, or $H = \frac{L^2}{D}$. Therefore $v = \sqrt{2g\frac{L^2}{D}}$, which is proportional to L, the length of the chord.

Again equating (192) and (186), two expressions for the final velocity v, we have $\sqrt{2gH} = g\frac{H}{L}t$, whence $t = \sqrt{\frac{2L^2}{gH}}$. But $L^2 = H \times D$, whence $t = \sqrt{\frac{2HD}{gH}} = \sqrt{\frac{2D}{g}}$, a quantity independent of L and H. This value of t is also the time of falling through the vertical diameter D. For, equating (177) and (179), $gt = \sqrt{2gh}$, whence $t = \sqrt{\frac{2h}{g}}$. If for h we substitute D, the diameter, we have $t = \sqrt{\frac{2D}{g}}$, the time of falling through that diameter which is equal to the value already obtained for the time of descent down the chord DC.

The same method of proof can evidently be applied to any chord AD drawn from the upper extremity of the diameter AC, hence the proposition is general.

265. Properties of Cycloid. If a body be supposed to descend through a given vertical distance AB, Fig. 142, by rolling over the paths AC, ADC, AEC, AFC successively, it is evident that while the velocity acquired on reaching C would be the same in all cases, the time occupied in making the descent would vary. Hence the question rises, over what path would the body descend most rapidly? or, in other words, what is the curve of swiftest descent? It was first demonstrated by J. Bernouilli, that the curve known as the *cycloid*,[1] possesses this property, so that if ADC be a semi-cycloid having a horizontal base, the body will reach C sooner than by any other path. Hence the cycloid is often called the *brachistochrone*.

Another curious property of the cycloid is that the time required to descend to C from all points on the curve ADC is the same. Thus the time of descending DC will be the same as that occupied in describing ADC, as in the latter case, the greater steepness of the arc about A compensates for the increase in the distance traversed. This important truth was discovered by Huyghens.

CHAPTER XII.

Curvilinear Motion.

266. Origin of Curvilinear Motion. Any moving body, if left to itself, continues its motion in a straight line because of its inertia. If any change occurs in the direction of the motion, it must evidently be due to some new force, whose line of action is inclined to the path of the body; and for each succes-

[1] The cycloid is the curve described by a point in the circumference of a circle rolling upon a straight line. Thus let DE, Fig. 143, be a circle, and P a point in its circumference. If the circle rolls over the straight line AB, the point P will describe the cycloid APB. AB is called the base of the cycloid, and from the mode of generation of the curve is evidently equal to the circumference of the generating circle.

sive change of direction a separate impulse of the deflecting force is necessary. A curve changes its direction at every point, therefore, in curvilinear motion, the number of impulses of the deflecting force in a unit of time must be infinite; that is, the deflecting force must act constantly. For example, a body moving in the line *AB*, Fig. 143, if uninfluenced by any additional force, will move onward in the same straight line towards *G*. If, however, when it arrives at *B*, it is acted upon by a force which alone would cause it to move over the line *BD* in a unit of time, according to the law of the composition of motions it will move in a new line *BM*, found by laying off *BK*, the distance it would traverse in a unit of time by virtue of its original motion, and drawing the diagonal *BM* of the completed parallelogram *BKMD*. The body will move in this line *BM*, until the deflecting force again acts at *C*, when it takes the direction *CN*. At each successive impulse of this force a change occurs in the path of the body, and if the number of these impulses in a unit of time is infinitely great, as is the case with a constant force, the broken line *ABCF* becomes a curve, the direction of which is constantly changing.

266. Definitions and Explanations. The deflecting force may be constant or varying in intensity. Its lines of action at different points of the curve may be parallel, or inclined to each other. If the lines of action at all points of the curve meet in a single point, this is called the *centre of force*, and the deflecting force is known as the *centripetal force*. The momentum of the body, which tends to carry it on in a straight line tangent to the curve, is called the *projectile* or *tangential force.*

A stone projected obliquely into the air furnishes an excellent example of centripetal force. The projectile force tends to carry it in a straight line, from which it is continually deflected by a force passing through the centre of the earth, and is thus caused to move in a curve.

267. Relation between Centripetal and Centrifugal Forces. *The centripetal force at any point of the curve is equal to that component of the projectile force which is directly opposed to it, and which tends to carry the body away from the centre of force.*

Let *PK*, Fig. 144, be an element of the curve. As the arc *PK* coincides with the chord *PK* for infinitesimal lengths, the body at a given instant may be considered as moving along that chord, which will therefore represent in magnitude and direction the resultant of the projectile and deflecting forces acting upon the body in question. Decompose this force into two others, represented by *PR*, tangent to the arc at *P*, and *PQ*, lying in the direction of the centre of force, both of which lines are sides of a parallelogram, of which *PK* is the diagonal. *PR* will represent the projectile, *PQ* the deflecting or centripetal force. Again, resolve *PR* into two components, PL directly opposite to PQ, and PK in the line of motion of the body, by constructing the parallelogram $PLRK$. The component PL tends to carry the body away from the centre, and is opposed to PQ. But as $PRKQ$ is a parallelogram, PQ and RK are equal; and as $PLRK$ is a parallelogram, PL is also equal to RK. Hence PL is equal to PQ, which represents the centripetal force.

That component of the projectile force, which is directly opposed to the centripetal force, is called the *centrifugal force.*

268. Additional Explanations. Curvilinear motion therefore arises from the simultaneous action of two forces, the projectile force, tending to carry the body onwards in a straight line, and the centripetal force, continually deflecting it from that line. The centrifugal force is not a power separate from the projectile force, but is only that component of it which is always opposite the centrifugal force. The projectile force is evidently the momentum of the body. When the revolving body is mechanically connected with the centre of force by a rod or string, the centripetal force is the tension of the particles of that rod or cord, caused by the centripetal force. In the case of free revolving bodies, as the planets, the centripetal force is generally the attraction of gravitation.

When a body revolves about a centre of force, the line joining that centre to any point of the path of the body is called a *radius vector*. The curve pursued by the body is known as its *orbit* or *trajectory*.

269. Law of Equal Areas. *If a particle revolves about a centre of force, the radius vector describes equal areas in equal times.*

Let us first suppose that the central force acts intermittently at equal intervals of time. Let the particle be at the point A, Fig. 145, moving with such velocity as would alone carry it to the point B in a unit of time, during which time the deflecting force alone would carry it to G. Under the united action of these forces the particle will move in the diagonal AC of the parallelogram $AGCB$. On arriving at C the particle tends to move over the line $CD = AC$ in a unit of time, because of its inertia. The deflecting force, however, now gives it such an impulse as alone would cause it to move over CG' (which may be greater or less than AG) in the same time. Impelled simultaneously by these two forces, the particle will move over the diagonal CE of the parallelogram $CG'ED$. Draw SD. The triangles SCE, SCD, are equal because SC and DE are parallel. The triangles SAC, SCD, are also equal, since $AC = CD$, and the vertex S is common to both. Therefore $SCE = SAC$. In like manner, SFE can be proved equal to SCE, and so on, for all parts of the trajectory. But the preceding reasoning holds, however short may be the interval between the successive impulses of the deflecting force; it is therefore true when that interval becomes infinitesimal, that is, when the force is constant. The lines AC, CE, EF, then become elements of the curve in which the body moves. Since, therefore, the equal elementary triangles SAC, SCE, SEF, are described in equal times, the same is true for any equal finite areas, as each of these contains the same number of elements.

270. Converse of preceding Proposition. *If there is a point within any orbit so situated that the line connecting it with the revolving particle describes equal areas in equal times, that point is the centre of force.*

Let $ACEF$, Fig. 145, be the orbit, and S a point so situated that the elementary areas SAC, SCE, SEF, described in successive elements of time are equal to each other. Then will S be the centre of the centripetal force acting upon the revolving body. Prolong AC, making $CD = AC$, and suppose the deflecting force to act intermittently, as in the preceding proposition. CD is evidently the space which would be traversed in the second element of time, were there no deviating force in action. But CE is the space actually traversed under the united action of both the projectile and deviating forces. Join DE, and complete the parallelogram $G'CDE$. CG', which is equal and parallel to DE, evidently represents the direction and magnitude of the deflecting force. To demonstrate the present proposition it is merely necessary to prove that the direction of this deflecting force is toward S, that is, that CG' lies in CS. Now the triangles SAC, SCE, are equal by hypothesis; the triangles SAC, SCD, are also equal, since they have equal bases, AC, CD, and a common vertex, S. Hence $SCD = SCE$, and as these triangles have the same base, SC, their vertices must lie in a line parallel to SC, that is, DE is parallel to SC, and hence CG' lies in SC. In the same manner it can be proved the direction of the deflecting force at all other points of the orbit, as E, F, etc., is directed toward S, hence S is the centre of force.

The two preceding propositions, together with the three following are extremely important from an astronomical point of view, as will be more fully seen in the chapter upon *Gravitation.*

271. Velocity at different Points of Orbit. The demonstration given in § 269 furnishes a simple method of comparing the relative velocity of a body at different points of its orbit.

Let v, v', be the velocities possessed by a particle at two points of its orbit, P, P', Fig. 146, and let SPQ, $SP'Q'$, be equal elementary areas described by the radius vector in an infinitely short time. The arcs PQ, $P'Q'$ being elements of the curve, may be considered as being traversed with uniform velocities, v, v', in time, t, and $PQ = vt$, $P'Q' = v't$. Draw PT, $P'T'$, tangents to the curve at the points P, P', and ST, ST', perpendiculars let fall upon those tangents from S. Denote ST by A, ST' by A'. The triangles SPQ, $SP'Q'$, are equal to $vt \times A$, $v't \times A'$, respectively. Hence as $SPQ = SP'Q'$, $v \times A = v' \times A'$, whence

$$v : v' :: A' : A. \quad (194.)$$

Hence the velocities at different points of an orbit are inversely proportional to the perpendiculars let fall from the centre of force upon the tangents to the orbit drawn from those points.

272. Law of Force in Orbit. The centripetal force in different orbits, and at different points of the same orbit, is proportional to the deflection of the curve corresponding to arcs described in equal infinitely short times, since the deflection is wholly caused by that force, which may be considered as constant during the time of describing elementary arcs. We are thus furnished with a method of comparing central forces, for if we know the form of an orbit, and the position of the centre of force, we can readily compute the deflections at any point, and thence the relative values of the centripetal force.

273. Elliptical Orbit. The most interesting application of this principle is the case of a particle revolving in an elliptical orbit about a centre of force situated at one of the foci.

Let S, Fig. 147, be that focus of the ellipse in which the centre of force is located, and SPQ, $SP'Q'$, equal areas described in an infinitely short time. The deflections RQ, $R'Q'$, parallel to the radii-vectores, SP, SP', measure the intensity of the centripetal force at the points P, P'. But by a property of the ellipse $RQ : R'Q' :: SP'^2 : SP^2$ (195), that is, *when the orbit is an ellipse, with the centre of force at one of the foci, the centripetal force at different points of the orbit is inversely proportional to the square of the distance of the revolving particle from the focus.*

The same law of force can be shown to hold when the particle describes either a parabola or a hyperbola.

274. Centripetal Force in Circular Orbit. The curvature of the circle being the same for every portion of the curve, the centripetal force in such an orbit is of constant intensity.

To determine its value, let C, Fig. 148, be the centre of the circle in whose circumference the particle revolves, PQ an arc described in an infinitesimal time, PR the space over which the body would pass in that time in virtue of the projectile force, $PM = RQ$, the space over which it would pass in virtue of the centripetal force. Call v the velocity of the particle, t the time of describing the arc PQ, and R the radius of the circle.

The arc PQ and the chord PQ will coincide, as the arc is an element of the circumference, and $PQ = vt$. By a property of the circle, $PQ^2 = PM \times PD$, or $v^2t^2 = PM \times 2R$, whence $PM = \frac{v^2t^2}{2R}$. Call f the acceleration which the centripetal force is capable of generating in the particle in a unit of time. Then, by the laws of uniformly accelerated motion, $PM = \frac{1}{2}ft^2$. Equating this with the value of PM previously found, $\frac{1}{2}ft^2 = \frac{v^2t^2}{2R}$, whence $f = \frac{v^2}{R}$. (196.) That is, *the acceleration which the centripetal force is capable of producing by its uniform action during a unit of time is equal to the square of the velocity of revolution divided by the radius of the orbit.*

Knowing this acceleration we can easily deduce the value of the centripetal force in units of pressure, or kilogrammes. Let F be the centripetal force thus estimated, and W the weight of the body. Then, since forces are proportional to their accelerations, $F : W :: f : g$, whence $F = W\frac{f}{g}$. Substituting the value of f, given in (196), we have, $F = \frac{W}{g}\frac{v^2}{R}$ (197), or for $\frac{W}{g}$ substituting its value, M, as given in equation (8), we have $F = M\frac{v^2}{R}$. (198.)

The centripetal force expressed in units of pressure is therefore equal to *the mass of the particle into the square of its velocity, divided by the radius of the orbit.*

In formula (198) if M is made equal to $\frac{W}{g}$, the force F is given in gravitation units. If M be made equal to the mass of the body in absolute units of mass, *i.e.*, in grammes or pounds, the result is expressed in absolute units of force. For $\frac{W}{g}\frac{v^2}{R}$ being the number of gravitation units, $\frac{W}{g}\frac{v^2}{R} \times g = W\frac{v^2}{R}$, is the number of absolute units of force in F (§ 50, p. 37), in which expression we may write M, the mass in absolute units, instead of W.

The centripetal force of any extended body, or system of bodies, can be shown to be the same as if the total mass of that body or system were concentrated at its centre of gravity. Hence the propositions which have been demonstrated as true for particles, can be applied to extended masses by simply considering the whole mass thus concentrated.

275. Different Expression for Preceding Formulas. Formula (198) can be put in another form, which is often of convenience. Let T be the time of revolution in the orbit. Then $vT = 2\pi R$, and $v = \frac{2\pi R}{T}$. Substituting this value of v in (198), we have $F = M\frac{4\pi^2 R}{T^2}$. (199.) Therefore *for bodies of the* *same mass revolving in circular orbits of different radii, the centripetal force is proportional to the radius of the orbit, and inversely as the square of the time of revolution.*

276. Velocity in Orbit. *The uniform velocity with which a body revolves in a circular orbit is equal to that which the centripetal force would generate by its uniform action upon a body falling through half the radius of the orbit.*

For by (196) $f = \frac{v^2}{R}$, and $v = \sqrt{fR}$. Suppose the centripetal force to impel the body from a state of rest until it attains a velocity v. Call S the space described. Then from (172), $v = \sqrt{2fs}$. We have, therefore, $\sqrt{fR} = \sqrt{2fs}$, whence $s = \frac{R}{2}$.

277. Body revolving in Vertical Orbit. If a body held by a cord be made to revolve in a vertical circular orbit, it is evident that the tension of the cord will vary from $F - W$ to $F + W$, (F being the centrifugal force, and W the weight of the body) according as the body is at its uppermost or lowest position. In order that a body may revolve in such an orbit, it is necessary that the centrifugal force shall be at least equal to the weight of the body, as otherwise the uppermost point of the curve could not be passed. It is easy to calculate from this the minimum velocity which must be given to the body to enable it to perform its revolution. For, with this minimum velocity, $F = W$. Hence $W = \frac{W}{g}\frac{v^2}{r}$, and $v = \sqrt{gR}$. (200.)

278. Experimental Illustrations. The action of the centripetal and centrifugal forces may be illustrated by the *Whirling Table*, one form of which is shown in Fig. 149. It consists essentially of a wheel A, to which a rapid rotation can be imparted by means of the larger wheel B connected with it by a cord or train of wheel work. If a frame EF (Fig. 150) be attached to the small wheel so as to revolve about the vertical axis A, a body W moving freely u on a horizontal wire, will show the action of the centrifugal force by moving towards D. The experiment may be made quantitative, and the laws of central forces verified, by affixing to W a cord passing over the pulleys B, C, as represented in the figure, and to which a weight P is attached which can be varied at will. The table is whirled until P just begins to rise because of the centrifugal force exerted from W by means of the cord. The velocity of rotation is noted, as well as the magnitudes of P and W, which data furnish a means of verifying formula (198.)

Many interesting experiments may be performed with the whirling-table, Fig. 151 represents a frame CD, containing two inclined tubes proceeding from a reservoir filled with water. On rotating this, the centrifugal force generated, causes the water to rise in the tubes. If a ball of iron, S, is placed in one tube, and a ball of cork, T, in the other, the ball S, being heavier than the water, will be thrown to the farther extremity of the tube on rotating the machine, while the lighter ball of cork T, will remain at the bottom, as the greater mass of the water causes it to possess a greater amount of centrifugal force.

If a glass of water be rotated rapidly in a horizontal plane, the liquid is elevated about the borders and depressed at the centre, the depression having a paraboloidal form. A flexible hoop of metal fixed at the point A

upon a vertical axis AB, and free at C, bulges out at its equator on being rotated, and assumes the form represented by the dotted lines. All the particles of the hoop tend to move as far as possible from the axis, in virtue of their centrifugal force, and this being greatest at those points having the greatest radii of rotation, the particles at the equator press so strongly outward as to cause the hoop to become elliptical.

The action of the centrifugal force can also be shown, and its laws verified, by suspending a heavy ball from a delicate spring balance, as in the figure (Fig. 153), and allowing it to swing as a pendulum. The excess of the index reading when the ball is vibrating and at its lowest point, over the reading when it is at rest, gives the centrifugal force directly. By varying the weight of the ball and the length of the sustaining cord, formula (198) may be illustrated experimentally.

279. Practical Applications and Illustrations. A practical application of the foregoing principles is illustrated by the machine often used in laundries for drying cloths, and known as the *hydro-extractor*. It consists of a large annular trough having its exterior surface perforated with holes, or made of coarse wire cloth. The wet cloths are wrung partially dry, and then placed in the trough, to which a rotation of from 1200 to 1500 turns per minute is then imparted. The water flies away from the cloth in virtue of the centrifugal force generated by the rotation, rendering it almost dry in a few minutes. A similar machine is used in England in the operation of sugar-refining. The syrup being boiled in vacuo, crystals of sugar form throughout its mass. To separate the molasses from these, the whole mass is poured into the trough of a machine resembling that already described. When this is revolved rapidly, the molasses flies off through the apertures, leaving a clear mass of soluble sugar behind. A small jet of steam directed against the outside of the trough prevents the expelled syrup from collecting there in a coating, and obstructing the exit of the remainder. The same process has also been applied quite recently, to separate the liquid colors used in printing from the more solid matters with which they have been mixed. These colors are usually very difficult to strain, but the present process seems to have rendered the operation quite easy.

An interesting illustration of centrifugal force occurs in the process of making crown-glass. The workman first obtains a quantity of melted glass upon the extremity of a long iron tube, which he shapes into a bell-like form. Then heating this again until the glass is softened, he places the tube horizontially upon an iron bar, and rotates it with great rapidity. The centrifugal force causes the bell to bulge out, and finally to become almost flat and of uniform thickness except at the very centre, where the instrument is attached, and from it small sheets of extremely brilliant and clear glass may be cut.

The *fanning-mill*, or *blower*, used for creating a blast of air, and applied to cleaning grain, the ventilation of buildings, etc., is an additional example of the application of the laws of centrifugal force. It consists of a drum MM, Fig. 154, within which is an axis carrying several paddles or vanes, N, N, is made to revolve rapidly. The centrifugal force produced in the air, carried round by the fans, causes it to rush through the tangential exit tube S, in a constant stream. Fresh air is drawn in through a large aperture near the axis of rotation.

When the rotation of a body becomes exceedingly rapid, the centrifugal force generated may be sufficient to overcome the cohesion of the particles composing it. Large millstones and fly-wheels, running at a high rate of speed, have been known to burst, the fragments scattering in all directions, and causing great destruction.[1]

280. Vehicle on Curved Road. When a carriage moves rapidly upon a curved road, the tendency of the centrifugal force is to overturn it. This will be understood by a reference to Fig. 155. The vehicle is acted upon by two forces, its weight, along the line GW, and the centrifugal force along GF. Suppose GW, GF, to be made proportional to these forces; then GR will represent their resultant, and if that line falls without the base formed by the wheels, it is clear that the vehicle will be overthrown. As the centrifugal force is proportional to $\frac{v^2}{R}$, the greater the velocity and the less the radius of the curve, the greater the tendency to overturn.

281. Depression of Inner Rail on Curve. In the case of railroad trains, the velocity is so great as to render it necessary to counteract this effect of the centrifugal force. This is done by making the inner rail lower than the outer one. That the stability may be unaltered by the motion in the curve, the resultant GR, Fig. 156, of the weight of the vehicle and the centrifugal force must fall midway between the wheels. To calculate the inclination requisite for this, let GW represent the weight of the car, and GF the centrifugal force. The slope of the road AB must be such that it is perpendicular to GR. But calling θ the inclination, tang $\theta = \frac{RW}{GW} = \frac{F}{W}$, or as $F = \frac{W}{g}\frac{v^2}{R}$, tang $\theta = \frac{v^2}{gR}$. (201.)

To find the linear elevation necessary to make tang θ equal to this quantity, call $h = BC$, the elevation required, d the distance AB between the rails. As θ is very small, AB may be considered as equal to AC, in which case tang $\theta = \frac{h}{d}$. Hence $\frac{h}{d} = \frac{v^2}{gR}$, and $h = \frac{dv^2}{gR}$. (202.) In the formula v is taken as the ordinary running velocity of the trains.

282. Centrifugal Force at Equator of Earth. The rotation of the earth on its axis, carries all bodies resting upon its surface in parallel circles, thus generating a certain amount of centrifugal force. The velocity being greatest at the equator, the centrifugal force is greatest there. To calculate its amount, we must first find the velocity of a body situated at the equator, and substitute this together with the proper values of g and R in equation (197). We have $F = \frac{W}{gR} \times \left(\frac{2\pi R}{t}\right)^2$, in which t is the time of a single revolution, or 86,164 seconds. Assuming $R = 6{,}377{,}300$ m., the mean equatorial radius, and $g = 9.8087$ m., $t = \frac{W}{9.8087 \times 6{,}377{,}300} \times$

$$\left(\frac{2 \times 3.14159 \times 6{,}377{,}300}{86{,}164}\right)^2 = W \times \frac{1}{289}.$$

The centrifugal force acting upon a body at the equator, is therefore about $\frac{1}{289}$ the weight of the body, and as on the equator, this weight and the centrifugal force are directly opposed, the weight of a mass is diminished in that ratio.

If the earth's velocity of rotation were to increase, the centrifugal force would grow greater and greater, till a point would be reached, when it

[1] For an account of a very destructive accident of this kind, see *Journal of the Franklin Institute*, Vol. LXV., p. 86.

would exactly counterbalance the force of gravity, and hence a body at the equator would lose all its weight. The velocity required can readily be found. That this may occur, we must have $W = \frac{W}{g}\frac{v^2}{R}$, or $v = \sqrt{gR}$, whence $v = 79.09$ m., or about 17 times their present velocity.

283. Variation of Centrifugal Force with Latitude. Let F denote the centrifugal force of a body situated at the equator, and F' that of a body at E (Fig. 157), in latitude θ. Since the revolution of the bodies about CD is accomplished in equal times, the centrifugal forces are proportional to the radii (199), that is, $F : F' :: CR : DE$. But $DE = CE \cos\theta = CR \cos\theta$. Hence $F : F' :: CR : CR \cos\theta :: 1 : \cos\theta$ (203), that is, the centrifugal force at any place is proportional to the cosine of the latitude of that place.

284. Effect of Centrifugal Force on Figure of Earth. The force F', represented by EG, Fig. 157, can be resolved into two components, represented by EH, EK, the former of which is perpendicular to the surface, the other tangential to it. The tendency of the tangential component upon any particle at E, is to force it towards the equator, and if the surface is supposed to be covered with a liquid whose particles are free to move, they will be driven onwards, accumulating about the equator, so that the mass will no longer be spherical, but will assume the form of an oblate spheroid. This accumulation will go on until the resultant AR (Fig. 158), of AT, the centrifugal force, and AM, the force of gravity, is normal to the surface of the figure, when the whole force acting upon the particle is exerted to push it against that surface, and the particle will evidently remain at rest. The form thus assumed is called the *spheroid of equilibrium*. Geology shows that the earth was once in a fluid state, in which case the spheroidal form would necessarily be assumed. Knowing the velocity of rotation, and the dimensions and weight of the earth, the amount of flattening can be calculated upon mathematical principles. This subject has been investigated by Huyghens, Newton and Maclaurin, whose results agree, in general, with those given by actual measurement, although from another cause the figure of the earth deviates slightly from that of a spheroid.

The oblate spheroid is not the only form of equilibrium which can exist in the case of a rotating liquid mass. The general problem of determining the possible forms of equilibrium in such a mass is extremely difficult, and its solution has scarcely been attempted. Jacobi has shown, however,[1] that an ellipsoid of three unequal axes, the least of which is the rotation-axis, is also a form compatible with equilibrium.

285. Plateau's Experiment. The effect of the centrifugal force in producing the spheroidal form of the earth can be illustrated experimentally by a method devised by Plateau. A small quantity of oil is poured into a mixture of alcohol and water, these liquids being combined in such proportions that the density of the mixture is just equal to that of the oil, which is therefore relieved from the action of gravity, and has no tendency either to sink or to rise to the surface. It can easily be collected in a single mass, which assumes the spherical form because of the internal molecular attraction acting among its particles. If now a vertical wire bearing a small disc be passed downward through the oil till the disc is at the centre of the sphere, and then made to revolve rapidly, the oil begins to revolve with it.

[1] See Thompson and Tait's *Treatise on Natural Philosophy*, Vol. I., p. 617.

As the velocity is increased the spheroidal form is assumed, the bulging out at the equator becoming more and more prominent. Finally, when the velocity becomes sufficiently great, a ring of oil is detached from the spheroid, and though entirely separated from the former, continues to revolve around it.

CHAPTER XIII.

WORK AND ITS MEASURE.—DYNAMICS OF RIGID BODIES.

286. Nature of Work. Work is performed whenever a force produces motion in opposition to a resistance; as when a weight is raised in opposition to the action of gravity, when a force compresses a spring, or moves a body against the resistance of friction, or the resistance of the air. It is proportional (1) to the amount of the resistance, and (2) to the distance through which that resistance is overcome; that is, to the distance through which its point of application is moved. Thus if a body be lifted, there is work done, which is greater in proportion as the mass is greater, and as the distance through which it is raised is greater.

Since the work done varies directly as each of these quantities, it varies directly as their product, and by choosing a suitable unit, the work performed may be said to be equal to the product of the resisting force into the distance through which it is moved. That is, if we represent the resistance by P, and the distance through which it is moved by S, the work E will be $E = PS$. (204.)

For example, if a weight of 100 kilogrammes be raised through 10 metres, and a weight of 10 kgrs. through 5 metres, the quantities of work performed in the two cases are to each other as 1000 to 50.

Both these elements, motion and pressure, are necessary to the performance of work. There is no work done when a force merely exerts pressure against a surface without causing motion, as when a mass is supported upon a prop. Neither is there any work done when the force has no resistance to overcome. Thus a body moving in free space performs no work. It is only when a resistance is overcome through a certain space that work ensues.

287. Case of Oblique Force. If the point of application of the resistance does not move along the line of action of the force P represented by AC, Fig. 159, but is constrained to follow some other path as AB, the work done is still equal to the resistance overcome into the distance through which it is overcome, or if the force AC, the resistance AE, and the reaction AG are balanced, the work is equal to that component of the force which is directly opposed to the resistance, multiplied by the distance through which the point of application is moved. That is, the work done in the case of Fig. 159 is represented by $AB \times AC \cos \theta$, or $P \cos \theta \times AB$, which is equal to $P \times AM$, as $AB \cos \theta = AM$. Hence *the work done by an oblique force is equal to the magnitude of the force multiplied by the projection of the path of the point of application upon its line of action.*

288. Application to Virtual Velocities. It is evident that if the distance AB is supposed to be very small, AM is the *virtual velocity* of the force P, and $P \times AM$, which is called its *virtual moment*, must represent the elementary quantity of work done during the time corresponding to that displacement.

It follows from equation (85) p. 74, that the algebraic sum of the virtual moments of any number of component forces is equal to the virtual moment of their resultant. Hence when moving against a resistance, the work done by any number of component forces is equal to the work done by their resultant. There is therefore no gain of work by the use of any mechanical power, or *the quantity of work which can be performed by any force is a constant in whatever manner the force may be applied.*

289. Case of Variable Force. If the resisting force is variable, the path of its point of application may be supposed to be divided into a a number of equal parts so small in magnitude that the resistance may be considered as constant during the time occupied in traversing one of them. The elementary quantity of work done in passing over one of these spaces, will be equal to the resistance into that space, and the total quantity of work will be found by a summation of the elementary quantities by the processes of algebra or the calculus.

290. Unit of Work. The unit of work in practical use by engineers is the *kilogramme-metre* in French, and the *foot-pound* in English measures. The kilogramme-metre is the work done in moving a force equal to the weight of 1 kilogramme through the space of 1 metre. The *foot-pound* is the work done in moving a force equal to the weight of 1 pound through the space of 1 foot; neglecting the difference of g between London and Paris, the kilogramme-metre is equal to 7.2331 foot-pounds.

In purely scientific investigations, however, these units are not employed. In such cases the unit of work is the work done in moving an absolute unit of force through a unit of space. Hence the absolute unit of work in French measure is the work done in moving a force equal to the weight of $\frac{1}{9.8087}$ grammes at Paris through 1 metre, and in English measure the work done in moving a force equal to the weight of $\frac{1}{32.1889}$ pounds at London over 1 foot.

291. Rate of Work. The amount of work performed depends simply upon the force overcome, and the distance through which it is overcome, and is totally independent of the time occupied in its performance. In other words, the work done is the same, whether it be performed more or less rapidly. In practical calculation, however, the element of time enters, and it is necessary to determine the quantity of work done in a given time. The *rate of work* is the work done in a unit of time, and is found by dividing the total work performed by the time occupied in its performance. That is, if PS be the work done in T seconds, $R = \frac{PS}{T}$ (205), is the rate of work. Thus, if a force move a resistance of 10 units over 100 metres in 5 seconds, the rate of the work, that is, the quantity performed in 1 second is $\frac{PS}{T} = \frac{10 \times 100}{5} = 200$ units.

The standard in ordinary practical use for comparing the rates of work of different motors, as steam-engines, water-wheels and the like, is the *horse-power*, which is a force capable of performing a work of 550 foot-pounds, or 76.0375 kilogramme-metres per second. A horse-power is therefore a force capable of raising 76.0375 kgrs., 1 m. per second, or 550 lbs., 1 ft. in that time.

To illustrate the application of this standard, suppose that a steamer moves against a resistance of 10.000 kgr. at a rate of 10 km. per hour. It is required to find the horse-power of an engine that will propel it with this velocity. The work done per second is 27778 kilogramme-metres, hence the required horse-power is $\frac{27778}{76 \cdot 0375} = 365.33$ horse-power.

292. Accumulated Work. Equation (204) gives the work done by a force when the resistance overcome, and the space through which it is overcome are known. But it frequently happens that this distance, S, is not given, but instead of it the velocity, V, with which the body is moving, is known. It is therefore important to find an expression for the work which can be done by the body in question, in terms of its mass and velocity. This can readily be found by ascertaining the space over which the body would move with a uniformly retarded motion before its velocity would be reduced to zero, and substituting this space as expressed in terms of V in equation (204).

If the body is moving with a velocity V, against a resistance P, which is capable of producing a retardation a, the distance S, over which the body will move, is $S = \frac{V^2}{2a}$. Substituting this value of S in (204), we have, $E = PS = P\frac{V^2}{2a}$. But $\frac{P}{a} = \frac{W}{g} = M$, whence $E = \frac{1}{2}MV^2$ (206).

The work which a moving body is capable of performing is therefore equal to half the mass into the square of its velocity.

If M is made numerically equal to $\frac{W}{g}$, as in the first system of the measurement of forces, the resulting value of E is expressed in foot-pounds or kilogramme-metres. If M is made equal to the weight of the body in grammes or pounds, as in the second system, the work E is expressed in absolute units.

The quantity $\frac{1}{2}MV^2$ is known as the vis-viva, or *living force* of the body.[1] Work thus stored up in a body is often called *accumulated work.*

To illustrate this by an example, let it be required to find the amount of work in kilogramme-metres which must be exerted to bring to rest a cannon-ball weighing 15 kgrs., and moving with a velocity of 400 m. per second. The work required must equal the vis-viva of the ball, that is, $E = \frac{1}{2} MV^2 = \frac{1}{2} \times \frac{15}{9.8087} \times (400)^2 = 122,340$ kilogramme-metres.

293. Energy.—Kinetic Energy. *Energy* is the power of doing work, and is either *Kinetic* or *Potential.*

Kinetic energy is the power possessed by a body of doing work in virtue of its force of motion, and is represented by the equation (206), $E = \frac{1}{2}MV^2$, in which E is the kinetic energy. The term kinetic energy was first introduced by Thomson and Tait. The term *actual* energy has also been applied to this kind of energy by Rankine.

If an unbalanced force acts upon a mass in motion, it is consumed in adding to the kinetic energy of that mass by continually increasing its velocity. The increment of energy during a given time, that is, the work stored up, is $K = \frac{1}{2}M(V_1^2 - V^2)$ (207), in which V, V_1, are the velocities at the beginning and end of the time considered. If the mass in motion moves against a resistance it evidently does work, which work is measured by the decrease of kinetic energy, that is, by $K = \frac{1}{2}M(V^2 - V_1^2)$ (208).

The storing up of energy may be illustrated by reference to the case of a cannon ball. By the continued action of the confined gases generated by the burning powder, the ball is moved along the bore of the gun with a continually accelerated motion, thus storing up a large amount of work, or, in other language, becoming possessed of a great amount of kinetic energy. This is gradually diminished by the resistance of the air, and finally, on striking any obstacle, as a wall, or an iron armor-plate, the ball penetrates until it has expended all its energy, when it comes to rest. The total work done is expressed by equation (206). Or, again, a stream of water, by running down its bed, acquires a considerable velocity, thus generating kinetic energy, which may be employed in running a water-wheel against which it impinges, and so carry machinery. The amount of work which this machinery can do in grinding corn, or rolling iron, or in any other species of work, is (after deducting the loss by friction, and in other ways) exactly equal to the kinetic energy of the moving water. The same is true of a windmill moved by air, and of any other form of motor.

[1] Vis-viva has generally been defined as equal to MV^2, but it is more simple to consider it as identical with the work which the mass is capable of performing.

294. Potential Energy. This kind of energy is the power of doing work possessed by a mass in virtue of its position. If a body is so situated that it is acted upon by a force which will produce motion in it, as soon as some restraining force is removed, and thence generate kinetic energy, it is said to have potential energy. Thus, a weight suspended at an elevation will fall as soon as the cord sustaining it is cut. The potential energy of a mass is expressed by the amount of kinetic energy which would be developed in it under the unimpeded action of the forces soliciting it. For example, the potential energy with regard to the surface of the earth, of such a body as we have supposed sustained at an elevation, is $E = PS$, if P is the weight, and S the elevation, or $E = \frac{1}{2}MV^2$, if V is the velocity which would be developed by the motion of the body until stopped by coming in contact with the earth. As other illustrations of potential energy, we may mention the case of a stretched spring, which is capable of doing work in returning to its normal state, and that of a head of water, which is capable of doing work by the kinetic energy acquired when it is allowed to flow freely under the action of gravity. The term *potential energy* was first used by Rankine. Helmholtz had previously applied to it the title *sum of the tensions*, and Thomson had called it *statical energy*.

295. Conservation of Energy. An examination of the examples adduced will show that there is a close relation between these two kinds of energy. We have seen how potential is converted into kinetic energy. Let us now consider the converse case, in which kinetic is converted into potential energy. The most familiar instance which can be adduced is the case of a body thrown into the air with a certain velocity. It possesses a definite amount of kinetic energy, which will be expended in raising the mass against the action of gravity, and will carry it to a certain height. Suppose the body to be stopped while at this uppermost point of its course. The amount of kinetic energy then possessed is evidently zero, while the amount of potential energy is exactly equal to the kinetic energy which it had on leaving the earth, as the velocity which it would acquire by falling freely from the point now occupied is equal to that with which it was originally projected upward. It is clear that the amount of potential energy developed here is equal to the amount of kinetic energy consumed in developing it.

Next let us consider the case of the body after it has begun to descend. As it falls its kinetic energy becomes greater with the increase in its velocity, while its potential energy becomes less, since it is continually approaching the ground. It will also be seen that for all points of its downward path the sum of its kinetic and potential energy must be a constant. For the potential energy at any point supposed will be that due to its distance from the ground, and the kinetic energy that due to its velocity acquired by falling from the highest point of its path. Hence, calling S the total height reached by the body, S' the height considered, $S - S'$ will be the distance of the body from the most elevated point reached by it. Now the kinetic energy is measured by the quantity $E' = P(S - S')$, and the potential energy by the quantity $E = PS'$; and the sum of these two expressions will be seen to be a constant, equal to PS. Or expressing the same fact by means of the equation for accumulated work, $E' = \frac{1}{2}MV'^2$ and $E = \frac{1}{2}M(V^2 - V'^2,)$ whence the sum of these is a constant equal to $\frac{1}{2}MV^2$. The same equations are evidently true for the ascending motion of the body.

A similar course of reasoning can evidently be applied to the case of a stretched spring, or of a head of water. Hence in all of these cases we see that the sum of the potential and kinetic energies possessed by a mass is a constant.

The preceding propositions are capable of being extended to a far more general form. *It can be shown that in any system not acted upon by external forces, and the masses composing which act upon each other along the line joining their centres of mass, with forces dependent only upon their mutual distance, and in no way upon their relative velocities, the sum of the potential and kinetic energies is a constant.*

This is a general statement of the fundamental principle of the *Conservation of Energy*. We shall see hereafter that there are various other forms of energy besides mechanical motion, into any of which such motion may be transformed.

The student will ask what becomes of the mechanical energy of a body stopped by friction, or the resistance of the air, or by being brought to rest on striking the ground. In all these cases kinetic energy disappears, and the body comes to rest, and yet there is apparently no potential energy developed in it. There is here a loss of mechanical motion, but it is transformed into another form of energy: viz., *heat*, which, as we shall see hereafter, is itself capable of developing mechanical energy under proper conditions, and the heat produced is an exact mechanical equivalent of the kinetic energy disappearing in the mass.

296. Energy of Rotating Bodies. Hitherto we have considered only the question of the accumulation of work in bodies having a translatory motion. In practice, however, it is frequently necessary to make use of the energy stored in a rotating body, as a fly-wheel. We proceed to consider the question of the determination of the kinetic energy of such a body.

297. Angular Velocity.—Angular Acceleration. In this investigation we shall make use of the *angular velocity* of the body, which is measured by the angle through which it rotates in a unit of time, and is the linear velocity of any point situated at a distance from the axis of rotation equal to unity.

Angular acceleration is the acceleration impressed in a unit of time upon such a point. The linear velocity of any point in a rotating body is equal to the angular velocity multiplied by its distance from the axis of rotation.

298. Determination of Energy. Let ABD, Fig. 160, be a body rotating about an axis through C, and at right angles to the plane of the paper. Suppose m to be the mass of a particle of the body situated at a distance r, from the axis of rotation. Let ω be the angular velocity of the mass. The kinetic energy, k, of the particle is $\frac{1}{2}mv^2$, or as $v = r\omega$, $k = \frac{1}{2}mr^2\omega^2$. The energy of any other particle of mass m', and situated at a distance r' from the axis, is $k' = {}_{2}m'r'^2\omega^2$. For other particles of masses, m'', m^n, situated with radii of rotation r'', r^n, $k'' = \frac{1}{2}m''r''^2\omega''^2$, $k^n = \frac{1}{2}m^n r^{n2}\omega^2$, and similar expressions may be obtained for all the particles of the body. The total kinetic energy of the body is E, which is equal to the sum of the energies of its particles, that is, $E = \frac{1}{2}mr^2\omega^2 + \frac{1}{2}m'r'^2\omega^2 + \frac{1}{2}m''r''^2\omega^2 + \frac{1}{2}m^n r^{n2}\omega^2$, or denoting by $\Sigma\, mr^2$ the sums of the products of the mass of each particle into the square of its radius of rotation,

$$E = \tfrac{1}{2}\omega^2\, \Sigma\, mr^2 \quad (209),$$

which expresses the amount of work which the body can perform before being brought to rest.

299. Moment of Inertia. The expression $\Sigma\, mr^2$ is of very frequent occurrence in treatises on dynamics, and is known as the *moment of inertia* of the body considered. We shall, in general, denote it by I. Hence $\Sigma\, mr^2 = I$ (210). Equation (209) may evidently be written $E = \frac{1}{2}\omega^2 I$ (211). *The kinetic energy of a rotating body is therefore equal to its moment of inertia multiplied by half the square of its angular velocity.*

The moment of inertia evidently varies with the form of the body, its mass and the position of the axis of rotation.

For continuous bodies equation (110) must be changed so as to consider the masses m, m', etc., as infinitesimal mass-elements, in which case we should have $I = \int r^2 dm$ (212).

300. Determination of Angular Velocity. Equation (211) furnishes a means of determining the angular velocity generated in a body by the expenditure of a given amount of energy. Thus suppose a cord, with a weight at one end, to be wound round a wheel, and to put this wheel in rotation by unwinding under the action of gravity. The angular velocity produced by the descent of the weight through a vertical height S, is to be determined. Let P be the magnitude of the weight, and I the moment of inertia of the wheel. The energy consumed in accelerating the wheel is PS, and the kinetic energy generated in the wheel is $\frac{1}{2}\omega^2 I$. Hence $PS = \frac{1}{2}\omega^2 I$, and $\omega = \sqrt{\frac{2PS}{I}}$, which is the angular velocity required.

301. Moment of Inertia about any Axis. If the moment of inertia of any body is known with regard to an axis passing through the centre of gravity, the moment of inertia relatively to any other axis can be found by means of the following demonstration.

The moment of inertia with respect to any axis is equal to its moment of inertia with respect to an axis passing through its centre of gravity plus the mass of the body into the square of the distance between the two axes.

Let AB, Fig. 161, be any body, G its centre of gravity, and C the point in which the axis of rotation cuts the plane of the paper, and let m be any element of mass. Call the distance CG between the two axes d, and denote the distances Gm, Cm, by l, r, respectively. Then by trigonometry,

$$Cm^2 = Gm^2 + CG^2 + 2GC \times Gm \cos CGm;$$

or, substituting for Cm, Gm, CG, etc., their values, and noticing that $\cos CGm = \cos PGm$, $r^2 = l^2 + d^2 + 2dl \cos PGm$. The moment of inertia of this particle relatively to C is $mr^2 = ml^2 + md^2 + 2dml \cos PGm$. As similar equations hold for each element of mass, we have

$$I = \Sigma mr^2 = \Sigma ml^2 + \Sigma md^2 + \Sigma 2dml \cos PGm,$$

or as d is a constant,

$$I = \Sigma mr^2 = \Sigma ml^2 + d^2\Sigma m = 2d\Sigma ml \cos PGm.$$

But $\Sigma m = M$, the mass of the body, and $ml \cos PGm$ is the statical moment of a particle relative to the centre of gravity G, since $l \cos PGm = GP$, hence $\Sigma ml \cos GPm = 0$. Also Σml^2 is the moment of inertia relative to G, which we will call I_0. Substituting these values in the preceding equation, we have

$$I = I_0 + Md^2. \quad (213.)$$

302. Radius of Gyration. The radius of gyration is a radius at whose extremity the whole mass of a rotating body may be supposed to be concentrated without altering its moment of inertia. That is, if ρ be that radius, and M the mass of the body, $M\rho^2 = I$, whence $\rho^2 = \frac{I}{M}$, and $\rho = \sqrt{\frac{I}{M}}$. (214). That is, the *radius of gyration of a body is equal to the square root of the quotient arising from dividing its moment of inertia by its mass.*

The following table, abridged from Rankine's *Applied Mechanics*,[1] gives the moment of inertia and radius of gyration of a number of bodies relative to an axis passing through the centre of gravity.

[1] Page 518.

TABLE.

Body.	*Axis.*	*Weight.*	I_0.	ρ_0^2.
Sphere of radius, r.	*Diameter.*	$\frac{4}{3}\pi wr^3$*	$\frac{8}{15}\pi wr^3$	$\frac{2}{5}r^2$
Spheroid of revolution—polar-semi-axis, a, equatorial radius, r.	*Polar Axis.*	$\frac{4}{3}\pi war^2$	$\frac{8}{15}\pi war^4$	$\frac{2}{5}r^2$
Circular cylinder—length 2a, radius r.	*Longitudinal Axis.*	$2\pi war^2$	πwar^2	$\frac{1}{2}r^2$
Do.	*Transverse Diameter.*	$2\pi war^2$	$\frac{1}{6}\pi war^2(3r^2 + 4a^2)$	$\frac{r^2}{4} + \frac{a^2}{3}$
Rectangular Prism—dimensions 2a, 2b, 2c.	*Axis 2a.*	$8wabc$	$\frac{8}{3}wabc(b^2 + c^2)$	$\frac{1}{3}(b^2 + c^2)$

303. Fly-Wheels. A knowledge of the subject of moments of inertia is of a great practical importance in the construction of all kinds of machinery which possess rotary motions. As a single example of the preceding principles, let us consider their application in the case of fly-wheels.

In all machines in which there is any variation in the magnitude, either of the impelling power, or of the resistance opposed, there must evidently be a fluctuation in speed, and it is generally the case in machines that during successive intervals of time, the variation in power applied and resistance offered is such that the energy received is alternately greater and less than the work to be performed. Now it is evident that when there is an excess of energy received over resistance overcome, this will be consumed in adding to the kinetic energy of the machine by increasing its velocity, and that when the energy applied is less than the resistance to be overcome, the kinetic energy of the machine diminishes, being consumed in overcoming the excess of the resistance offered over the impelling force applied at that moment, thus slackening the speed. This alternate increase and diminution of speed evidently varies according to the nature of the motive power, and the work performed. Let us call V, V', the maximum and minimum velocities of the *driving point*, or that part of the machine applied to overcome the resistance, in which case the fluctuation of speed will be represented by the quantity $V - V'$; and let V_0 be the mean velocity, equal to $\frac{V - V'}{V_0}$. The quotient $\frac{V - V'}{V_0} = C$ (215), will express the ratio of the fluctuation of velocity to the mean velocity, and is called the *coefficient of fluctuation of speed*. Now if we denote by W a mass which, if situated at the driving-point of the engine, would have the same actual energy as the entire machine, the maximum kinetic energy of the mass will be $\frac{W}{g}\frac{V^2}{2}$, and the minimum $\frac{W}{g}\frac{V'^2}{2}$, so that the variation

* w = weight of unit of volume.

in kinetic energy will be $Q = \frac{W}{g}\left(\frac{V^2 - V'^2}{2}\right)$. (216.) The mean kinetic energy is $M = \frac{W}{g}\frac{V_0^2}{2}$ (217), or as $V^0 = \frac{V + V'}{2}$,

$$M = \frac{W}{2g}\left(\frac{V + V'}{2}\right)^2. \quad (218.)$$

Hence, dividing (216) by (217), $\frac{Q}{2M} = \frac{V - V_1}{V_0} = C$, or substituting the value of M, given in (217), $C = \frac{V - V'}{V_0} = \frac{gQ}{WV_0^2}$. (219.) Hence the coefficient of fluctuation can evidently be ascertained by experimentally determining Q and V_0, if W is known.

Now in most machinery, if there were no regulating apparatus, this coefficient would be extremely great, as the variation of the relation between the effective force and resistance would render the difference between V and V' very large; the excess of the resistance over the energy applied might be sufficient to stop the machine, and would at any rate cause serious jarring. To obviate this, a *fly-wheel* can be attached to the rotating shaft of the engine. This is a wheel with a heavy rim, which possesses a great moment of inertia, and stores up energy when there is an excess of driving power, and gives it out to overcome the resistance when the driving power diminishes. It is clear that the coefficient of fluctuation will in this way be greatly diminished, as V, V', will become more nearly equal. In practice the fly-wheel is made of such dimensions as to reduce this coefficient to about $\frac{1}{32}$ for ordinary machinery, and $\frac{1}{50}$ to $\frac{1}{60}$ for delicate machinery.

The dimensions of the wheel in any given case can be ascertained as follows. Let $\frac{1}{n}$ be the coefficient of fluctuation which is to be obtained, and let the angular velocity with which the wheel is to revolve, as determined by the kind of work to be done, be denoted by ω. Then equation (219) becomes $\frac{1}{n} = \frac{gQ}{WV_0^2}$. If I be the moment of inertia of the fly-wheel necessary, and ω the proper angular velocity, $\frac{1}{2}\omega^2 I$ will be its kinetic energy, which may be taken as equal to the energy of the mass W, supposed to be concentrated at the driving-point. Hence $\frac{1}{2}\omega^2 I = \frac{1}{2}WV_0^2$, and $WV_0^2 = \omega^2 I$. Substituting this value in (219), $\frac{1}{n} = \frac{gQ}{\omega^2 I}$ (220), whence $I = \frac{ngQ}{\omega^2}$ (221), from which equation I can be determined.

Capt. Von Schuberszky, of the Russian engineers, has proposed to furnish heavy freight trains with large fly-wheels, placed on a truck immediately behind the locomotive, which will store up energy while descending inclines, and give it out as useful driving power when a level or ascent is reached, instead of wasting it by the use of brakes.[1]

304. Pressure upon Axis of Revolving Body. When a body revolves about an axis there is in general a pressure produced upon that axis, due to the centrifugal force of the particles composing the mass. The axis may, however, be so situated

[1] See *Rep. Commrs. to Paris Exposition*, Vol. III., p. 153.

that the centrifugal force due to any particle is just balanced by that exerted in the opposite direction by another particle, in which case there will evidently be no pressure upon the axis.

The effect of an unbalanced centrifugal force is twofold, tending (1) to shift the axis of rotation as a whole, *i.e.*, to produce a translatory motion of the axis; and (2) to change the angular position of the axis. To understand this let us consider Fig. 162, which represents a body *ADBC*, revolving about an axis *AB*. The portion *ADB*, lying upon the right of the axis, evidently tends to move it in the direction *ED*, and the portion *ACB*, lying to the left of the axis, tends to move it in the direction *EC*. But as the centrifugal force of *ADB* is greater than that of *ACB*, owing to its greater mass, and the greater distance of most of its particles from *AB*, there will be a resultant force, tending to shift the axis as a whole to some new position in space; in the case represented in Fig. 162, to a line at the right of *AB*, and parallel with it.

To see how the tendency to angular deviation is produced, let us notice the body *AGBH*, Fig. 163, which revolves about *AB* as an axis. Now the centrifugal forces of *AGDB*, *BHCA*, act in opposite directions, and as *AGDB* is greater than *BHCA*, there will be an unbalanced centrifugal force in the direction *ED*, as shown in the preceding case. Also since the mass of the body is not symmetrically distributed about *CD*, the centrifugal force of the portion *AGDE* minus that of *ACE*, can be represented by a single force *F*, applied at some point, as *S*, within *AGDE*, while the centrifugal force of *CHBE* minus that of *DEB*, may be represented by a single force *F''* applied at *T*. These two forces evidently act as a couple (*centrifugal couple*) to produce rotation about an axis at right-angles to the plane of *AB*, and hence tend to produce an angular deviation of that axis.

305. Free Axes and Principal Axes. A consideration of Figs. 162, 163, will show that if the axis of revolution passes through the centre of gravity of the body *ADBC*, the tendency to shifting it as a whole disappears, since the mass, and consequently the centrifugal force, is equally distributed on either side of the axis. That there may be no tendency to deviation of the axis, in which case the centrifugal couple must reduce to zero, the axis *AB* must also be an axis of symmetry of the body, since in this case the forces *F*, *F'*, will be directly opposed, as shown in Fig. 164.

It is evident that if the axis of revolution, *AB*, is parallel to an axis of symmetry, without passing through the centre of gravity, as shown in Fig. 162, there will be no tendency to angular deviation, though there is a tendency to translation of the axis.

Any axis about which a body may revolve without causing any tendency either to shift the axis as a whole, or to produce angular deviation, is called a *free axis* or *permanent axis*, because if the body is perfectly free to move, the axis will not change its position. It is proved by analysis that every body, or system of bodies, has at least three free axes, which are at right-angles to each other, and intersect at the centre of gravity of the body or system.

Any axis about which a body may revolve without causing any tendency to an angular deviation of the axis is called a *principal axis*. For each point of any body, or system of bodies, there are at least three principal axes at right-angles to each other. For points on either of the free axes, the principal axes are parallel to the free axes. The latter are evidently the principal axes passing through the centre of gravity. It can be shown that the moment of inertia of the body or system, relatively to one of its principal axes is greater, and relatively to another is less, than relatively to any other axis through the point considered.

In an ellipsoid having three unequal axes, the free axes are the three axes of that solid; in a right elliptical cylinder the axis of the cylinder, and the major and minor axes of the elliptical section of the cylinder at its middle point, are free axes; any diameter of a sphere is a free axis, also any diameter of the equator of an oblate or prolate speroid, together with its polar axis.

306. Practical Applications. In any rotating piece of machinery it is desirable that the resultant centrifugal force be reduced as much as possible, since the friction and strain of the machine are increased thereby, while the continual change in the direction of the centrifugal force as the moving body rotates, may give rise to dangerous jarring of the machine. Hence every rapid rotatory motion should, if possible, have its axis coincident with the free axis of the rotating mass.

The practical application of this is illustrated in the driving-wheels of locomotives. As the heavy cranks to which the connecting-rod is attached, would, if unbalanced, produce serious jolting of the locomotive as they revolve, the space between the opposite spokes of the wheel is filled up solid, that the axis of revolution may be as nearly as possible a free axis.

307. Axis of Stable Rotation. Although a body revolving accurately about any one of its free axes has no tendency to deviate from it, yet it is only in a condition of stable rotation when that axis is the one relatively to which the moment of inertia of the body is the greatest, which is, in general, the shortest axis. It is only in this case that it will return to its original axis if slightly displaced from it. To show this, suppose $CHDE$, Fig. 165, to be a body rotating about CD, the longest of its free axes. So long as the position of the body is undisturbed, the revolution will continue unchanged, but if it is slightly displaced in any way, so as to assume the position $C'H'D'E'$, Fig. 166, the rotation continuing about AB will generate a centrifugal couple $F - F'$, which tends to deviate the body still more from its original position, and which will disappear only when the shorter axis, $E'H'$, coincides with the axis of rotation. Hence the body is in stable equilibrium only when it is rotating about its shortest diameter, *i.e.*, about that one of its free axes with regard to which the moment of inertia is the greatest possible.

Thus a ring rotates in stable equilibriun only when revolving about an axis perpendicular to its plane; an ellipsoid of three unequal axes only when revolving about its shortest diameter, and an oblate spheroid when revolving about its polar axis.

308. Case of the Planets. It is worthy of notice that the oblate spheroid having its polar axis for an axis of rotation, the form naturally assumed by the planets, is one of the very few cases in which the axis of rotation is permanent. For in case of any external disturbance, producing a temporary swerving of the body, the tendency of the centrifugal force is to cause it to return to its original axis. Such perturbations are constantly acting upon every planet, and were not the variations caused by them confined within very narrow limits, they would lead to a continual change in the axis, and hence to a continual variation in the seasons, and in the distribution of land and water. Were the earth a prolate spheroid, for example, there would be no power to prevent the passing of the rotation-axis from one position to another, on the occurrence of any perturbation, and any suchchange, if once introduced, would go on forever.

309. Experimental Illustrations. The stable and unstable equilibrium of rotating bodies can be illustrated experimentally with the apparatus represented in Fig. 167. AB is a body suspended by a cord, by means of which it can be made to revolve very rapidly. If it be so hung as to rotate about any axis but its shortest one, the slight disturbance produced by the swaying of the string will displace it a little, and the body will then move until it assumes the position $A'B'$. A prolate spheroid, rotating about its longest axis, or a ring rotating about one of its diameters, shows this very clearly. An oblate spheroid made to revolve about one of its equatorial axes, will rise until it revolves about its polar axis, but if made to revolve originally about its polar axis, the equilibrium is seen to be stable.

310. Composition of Rotations. When a body is affected with two or more rotations about inclined axes, these rotations may be combined precisely as we combine couples. Thus suppose a body to be acted upon by two forces which separately would cause it to rotate about the axes AB, AC, Fig. 168, with angular-velocities a, a', respectively. To find the resultant axis of rotation, we may consider the rotating forces as forming two couples with inclined axes, and determine the magnitude and direction of the resultant couple, which will represent the resultant rotation in magnitude and direction. Hence we simply lay off on AB a distance Ab, proportional to a, and on AC a distance Ac proportional to a', and complete the parallelogram $Abdc$, the diagonal of which, Ad, represents the magnitude of the resultant rotation, and is also its axis. It is clear that in the same manner we may combine three or more rotations about axes lying in the same, or in different planes.

311. Gyroscope. The phenomena of the composition of rotations are excellently illustrated by means of the *gyroscope*. This instrument consists of a heavy, well balanced wheel, AB, Fig. 169, revolving with as little friction as possible about an axis DC. This axis is held in a ring, KH, to which is attached a projecting rod EF, with a pivot E, which may be placed in a socket G. If a rapid rotation be now communicated to the wheel AB by unwinding a cord from the axis CD, and the pivot E be rested in the socket G, with the axis horizontal, the instrument will not fall from the support as might be expected, but will remain nearly horizontal, and rotate slowly about E, in the direction indicated by the arrow.

To explain this action, let us consider the motions with which the wheel is affected. Suppose the axis of revolution to be so placed as to coincide with OX, Fig. 170. The wheel has a rotation about CD, Fig. 169, due to the original impulse. But as soon as E is placed in G, the weight of the apparatus, whose centre of gravity is then unsupported, gives it a tendency to fall, thus rotating about the point E, and in the plane of CD; that is, about an axis OY, Fig. 170, at right-angles to OX. Let the angular velocity of the rotation about OX be represented by Oa, that about OY by Ob. The direction and magnitude of the resultant rotation will be represented by Oc; hence this resultant motion must take place about the axis OX', and the metallic axis of the gyroscope (CD, Fig. 169) will move so as to approach this position, thus assuming a retrograde motion. But as the axis about which the rotation due to gravity takes place is always at right-angles to the axis of revolution of the wheel, OY will recede as the wheel moves towards it, and hence a continuous retrograde revolution will take place about OZ.

But the preceding construction is complete only for the first instant of the motion. For as soon as the movement about OZ has begun, there are evidently three simultaneous rotations to be compounded instead of two. To ascertain the effect of these, let us consider Fig. 171, in which Oc represents the rotation about the axis OX', Ob that about OY, and Oe that about OZ. The resultant rotation will be represented by Of, the diagonal of the parallelopiped $Ofce$, of which Oc, Ob, Oe, are adjacent edges. Hence the instrument will move so that its axis may assume this position. But as this movement takes place OY evidently recedes, and so the orbital rotation is continuous. Also since Of is inclined above the horizontal plane $X'OY$, the axis of the gyroscope tends to elevate itself. As Of is greater than either Oc or Ob, the angular orbital velocity would continually increase, were it not counterbalanced by friction and the resistance of the air. The effect of the orbital motion is to elevate the axis of the gyroscope, that of gravity to depress it; hence the axis rises or falls, as one or the other of these predominates, which latter circumstance is determined by the angular velocity of the wheel, and the position of the centre of gravity of the total mass of the gyroscope.

If the instrument is balanced about E, Fig. 169, by suspending a weight from F, so that the centre of gravity is over the pivot E, there is no tendency to rotation about OY, and hence no orbital revolution. If the weight at F is so great as to bring the centre of gravity of the instrument to the left of E, the gyroscope tends to rise, and the orbital motion becomes direct. The same effect occurs if the rotation of the wheel be reversed.

If the gyroscope be held in the hands and made to rotate rapidly, on attempting to incline it in one or another direction, a strong resistance to such change will be experienced by the hand.

The gyroscope also offers an experimental illustration of the astronomical phenomena of the constant parallelism of the earth's axis to itself, the precession of the equinoxes and nutation.

312. D'Alembert's Principle. When a rigid body is caused to move under the action of any external force, it generally occurs that some of its particles move slower, and some faster than they would move if there were no rigid connection between them. Suppose, for example, that we have two balls, one of gold and the other of some light substance, as pith. We shall see hereafter that these bodies would fall through a vacuum with equal velocities, but that the resistance of the atmosphere retards the pith more than the gold ball, so that in the air the descent of the latter will be more rapid than that of the former. Suppose now that the two are firmly connected. It is clear that they must now both fall with the same velocity, and that the gold ball moves less rapidly than if free. A portion of the impelling force impressed on the gold ball is not expressed in its momentum, while the momentum expressed in the pith is in excess of that impressed upon it, this difference arising from the connection of the bodies. Now the total expressed force must be equal to the total impressed force, as no energy can be lost; hence the pith ball must gain in momentum exactly as much as the gold ball loses. The same law evidently holds for all cases of motion in rigid bodies, whose particles tend to move with different velocities. Hence, in general, *the resultant of the impressed forces must be equal to the resultant of the expressed forces.*

This theorem is known as *D'Alembert's Principle*, from the mathematician who first enunciated it before the Academy of Sciences of Paris, in 1742.

313. Angular Acceleration of Body. Let AD, Fig. 172, be a body capable of moving about an axis through C, at right-angles to the plane of the paper, and suppose it to be acted upon by any force F, applied at a distance $BC = R$ from C. It is required to find the angular acceleration which the body assumes. Denote the angular acceleration, that is, the angular velocity generated in 1 second by ω, and consider first the case of a single particle P, of mass m, at a distance r from C. The velocity of P is $r\omega$, and its momentum $mr\omega$. The moment of this momentum relatively to C is $mr\omega \times r = mr^2\omega$. For other particles, m', m'', m^n, at distances r', r'', r^n, from the axis, similar expressions, $m'r'^2\omega$, $m''r''^2\omega$, $m^n r^{n2}\omega$, would be obtained, and the sum of these, or $\Sigma mr^2\omega$ is evidently the moment of the momentum of the whole body relatively to C. Now the force generating this momentum is the moment of F relatively to C, that is, $F \times R$, and as the total *impressed* force acting upon AD must equal the total force *expressed* in its motion, according to D'Alembert's Principle, we have $FR = \Sigma mr^2\omega$, or $FR = \omega \Sigma mr^2$ (222), as ω is a constant, whence $\omega = \dfrac{FR}{\Sigma Mmr^2}$ (223), or denoting Σmr^2, which is the moment of inertia of AD relatively to C, by I, $\omega = \dfrac{FR}{I}$. (224.)

If the force generating the motion is the weight of the body, acting through G, its centre of gravity, the impressed force will be $W \times GC = WR_0$, whence in the case supposed,

$$\omega = \frac{WR_0}{I} \quad (225).$$

314. Centre of Oscillation. The angular velocity which would, in the latter case, be assumed by any one of the particles of which the body AD is composed, would be found from the equation $\omega_p = \frac{w_p r_p}{I_p}$, in which ω_p is the angular acceleration which the particle tends to assume, w_p the weight of the particle, r_p its distance from C, and I_p its moment of inertia relatively to the axis of rotation. As $I_p = m_p r^2_p$, $\omega_p = \frac{w_p r_p}{m_p r^2_p} = \frac{g}{r_p}$ (226), whence it appears that the particles most distant from C tend to rotate more slowly than those nearer that point. The angular acceleration, ω, of the whole body, will therefore be greater than that which would be assumed by the particles farthest from C, were there no rigid connection, and less than that which would be assumed by the nearer particles. Hence there will be a point somewhere upon the line BC, which will possess the same angular acceleration as if it were not rigidly connected with the rest of the body. Let O be the position of that point. We wish to determine its distance, $CO = l$, from the axis through C, upon which it is suspended. The angular acceleration of O, were it free, would be $\frac{g}{l}$ (Eq. 226). This must equal ω, the angular acceleration of the body AD. That is, $\frac{g}{l} = \frac{WR_0}{I}$, or as $W = Mg$, $\frac{g}{l} = \frac{MgR_0}{I}$, whence $l = \frac{I}{MR_0}$ (227). *The point O is therefore at a distance from C equal to the moment of inertia of the body relatively to the axis through that point, divided by the statical moment of the weight relatively to the same axis.*

The point O is called the *centre of oscillation* of the suspended body AD.

In case the body vibrates under the influence of its weight, it performs its oscillations in the same time as if the whole mass were concentrated at O, since the angular accelerations in the two cases would always be the same.

315. Convertibility of Centres of Suspension and Oscillation. A proposition of great importance is the following. *The centres of suspension and oscillation are mutually convertible.*

If O is the centre of oscillation of a body suspended at C, and vibrating under the influence of its weight, then if the body be suspended from O its new centre of oscillation will be at C.

To demonstrate this, let I_0 be the moment of inertia of AD, Fig. 172, relatively to its centre of gravity, and I' relatively to the axis of suspension through C. Then denoting the distance from C to G by d, that from G to O by d', we have from Eq. (213), p. 145, $I' = I_0 + Md^2$. But $CO = l = \frac{I'}{Md} = \frac{I_0 + Md^2}{Md} = d + \frac{I_0}{Md}$. (228.) Suppose now that the body were suspended at O. Call l' the distance from the new axis of suspension O, to the new centre of oscillation, and I'' the moment of inertia of the body relatively to the axis through O. Then $l' = \frac{I''}{Md'} = \frac{I_0 + Md'^2}{Md'} = d' + \frac{I_0}{Md'}$ (229). Now from (228), as $l = d + d'$, $d + d' = d + \frac{I_0}{Md}$, whence $dd' = \frac{I_0}{M}$, and $d = \frac{I_0}{Md'}$. Substituting this value in (229) we have $l' = d' + d = l$ (230), whence it follows that the new centre of oscillation is at C, the former point of suspension.

316. Centre of Percussion.. The centre of oscillation can also be shown to be the *centre of percussion*, which is the point at which a body suspended from an axis may be struck by a blow in its plane of rotation without producing any pressure upon the axis; and, conversely, the point at which a body moving about an axis must strike an obstacle, that there may be no blow communicated to the axis.

A knowledge of the position of this point is practically applied by experienced batters, who learn by practice to strike the ball with that part of the bat in which the centre of oscillation of the system formed by the moving arm and bat, is situated. In this case, the whole force of the blow is spent in overcoming the motion of the ball, and no jar is sustained by the hand, otherwise an unpleasant stinging sensation is produced.

The centres of oscillation and percussion of a straight rod suspended at one extremity lie in the axis of the rod at a distance from the point of suspension equal to two-thirds the length. Hence if a stick held in the hand, and swung about the wrist, strikes an obstacle at that point, no blow will be felt by the hand, and at no other point will this be the case.

317. Axis of Spontaneous Rotation. If a body be impelled by a blow acting at its centre of gravity, it can be shown that the motion which it receives will be one of translation only; but if it be struck at any other point, the body will become affected with a double motion of translation and rotation, receiving the same motion of translation as if the blow had been delivered at its centre of gravity, and the same rotation as if the body had been suspended on an axis through the centre of gravity. From this it follows that if an impulse be communicated to a free body, the combined translation and rotation will cause its motion to be the same as if it revolved about some fixed axis, this axis changing, however, from moment to moment. The reason of this will be seen if it is remembered that the translatory motion is equally partaken of by all the particles, while the rotation causes some particles to move backwards, and others to move forwards, relatively to the direction of the translation, the rotatory velocity of any one particle being greater in proportion to its distance from the centre of gravity. If the backward motion of certain particles due to the rotation exceeds their forward motion due to the translation, they will evidently move backwards in space, and as the remaining particles move forward, there will be same point at which the backward motion ceases, and the forward begins. The resultant motion of the body is therefore the same as if it revolved about an axis passing through this point. The position of this axis changes as the body moves, because the position of the centre of gravity changes. That axis about which the body tends to revolve at the first instant of its motion, is called the *axis of spontaneous rotation*, and the successive axes about which such a body seems to rotate are called *instantaneous axes*. The position of the axis of spontaneous rotation evidently depends upon the point at which the body is struck, and it can be demonstrated that the axis of suspension corresponding to a given centre of oscillation is the axis of spontaneous rotation for a body struck at the latter point. Hence if a free body be struck at any point, it begins to move as if rotating about an axis so situated that if the body were suspended from that point, the point struck would be the corresponding centre of oscillation.

It will be seen that this fact affords an explanation of the principle already explained, that the centres of percussion and oscillation are identical, for if a blow be struck at the centre of oscillation, the axis of suspension will be the axis of spontaneous rotation, and hence no shock will be sustained by it.

References.

Treatise on Natural Philosphy, by Thomson and Tait. Chapter on *Work.*

Treatise on Infinitesimal Calculus, by Bartholomew Price; Vol. IV,. *Dynamics of Rigid Bodies.* (Oxford, 1872.)

Manual of Applied Mechanics, by W. J. M. Rankine; (4th Ed., London, 1868) Part V., *Dynamics.*

Manual of Machinery and Mill-work, by W. J. M. Rankine. (London, 1869.) Part II., *Dynamics of Machinery.*

Manual of the Mechanics of Engineering and of the Construction of Machines, by Julius Weisbach. Translated by E. B. Coxe. Vol. I., *Theoretical Mechanics*, Section V., *Dynamics of Rigid Bodies*, Chap. I., II. (New York, 1870.)

Correlation and Conservation of Forces. Essays by Grove, Helmholtz, Mayer, Liebig and Carpenter. Edited by E. L. Youmans. (New York, 1865.)

CHAPTER XIV.

GRAVITATION.

318. Law of Universal Gravitation. The force of gravity or weight, to which we have frequently had occasion to refer in preceding chapters, is only a particular case of the general law of universal gravitation, which we now proceed to consider. It is frequently known as *Newton's Law*, from its discoverer, and may be stated as follows: *Every particle of matter in the universe attracts every other particle with a force varying directly in the compound ratio of their masses, and inversely as the square of their distance.*

It follows from this that if we represent by ϕ the attraction of a unit of mass upon a unit of mass at a unit of distance, the mutual attraction, G, of two bodies of masses m, m', at a distance d is expressed by the formula $G = \phi \frac{mm'}{d^2}$ (231). For if ϕ be the attraction of one unit of mass upon another at a distance unity, the attraction of a unit of mass upon a mass m is ϕm, and the attraction of a mass m' upon m is $\phi mm'$. If the distance instead of being unity is equal to d, since gravitation varies inversely as the square of the distance, the attraction becomes $\phi \frac{mm'}{d^2}$.

319. Method of Proof of Law. This proposition was first enunciated by Newton, who showed it to be a direct consequence of the principles known as the *Three Laws of Kepler*, which that astronomer had established by astronomical observation. These are,

I. *All the planets move in elliptical orbits, of which the sun is in one focus.*

II. *The radius vector of any planet describes equal areas in equal times.*

III. *The squares of the times of revolution of the planets are proportional to the cubes of their mean distances from the sun.*

The following is an outline of the course of reasoning followed by Newton. In the first place the proposition demonstrated in § 270, p. 131, shows that since the radius vector of a planet describes equal areas in equal times, there is a deflecting force, which must pass through the centre of the sun, which last fact is also a consequence of the law of gravitation. For analysis proves that the law being true, the sum of the attractions exerted upon any external body by the particles of matter of a sphere (and approximately of a spheroid of small eccentricity), composed of concentric homogeneous layers, is the same as if the whole mass of the solid were concentrated at its centre, to which centre, therefore, any body subject to the attractive influence of the sphere would be drawn.

The law of the variation of the force in the inverse ratio of the square of the distance, follows from the ellipticity of the orbits, as shown in the proposition demonstrated in §, 273, p. 132. That portion of the law relating to the mass of the body also follows from the law of centripetal force in an elliptical orbit.

To show more clearly the mathematical reasoning employed, let us demonstrate the law in the simplest case, that of a planet moving in a circular orbit. As the centripetal or deflecting force in the case of the planet is the attraction of the sun, if we denote by G, G', the attractions exerted by the sun upon two planets at distances d, d', from its centre, are by § 275, p. 133, $G = M\frac{4\pi^2 d}{T^2}$, $G' = M'\frac{4\pi^2 d'}{T'^2}$, from which it follows that

$$G : G' :: M\frac{d}{T^2} : M'\frac{d'}{T'^2}.$$

But by Kepler's 3d Law, $T^2 : T'^2 :: d^3 : d'^3$. Hence

$$G : G' :: \frac{M}{d^2} : \frac{M'}{d'^2} \quad (232),$$

a proportion expressing Newton's Law. A similar demonstration can also be given in the case of the ellipse, parabola or hyperbola, so that under the influence of gravitation a body may move n either of these curves.

Newton next verified the law of inverse squares in the case of the earth and moon, showing also by his method of proof that the deflecting force acting on the planets is the same in nature as that which causes a body to fall to the ground.

Let AE, Fig. 172, be a portion of the moon's path, which it describes in 1 minute. Were it not for the deflecting action of its gravitation to the earth, it would move over the line AB in the same time. BE, which is sensibly equal to DE for small arcs, is therefore the amount of its deflection while traversing AE, that is, the amount by which it falls towards the earth in one minute. Knowing the dimensions of the moon's orbit, and the velocity of describing it, this distance can easily be computed, and compared with the space through which a body, at a distance from the centre of the earth equal to the moon's distance, would fall, assuming the law to be true. The deflection of the arc AE, from the tangent AB, is found to be 4.9 m. The moon therefore falls to the earth 4.9 m. in 1 minute. At the surface of the earth the space traversed by a falling body in 1 minute is 4.9×3600 m. Since forces are proportional to their accelerations, if we call G, G', the force of terrestrial gravitation at the earth's surface, and at the moon, and d, d', the radii of the earth and of the moon's orbit,

$$G : G' :: 4.9 \times 3600 : 4.9 :: 3600 : 1.$$

But $$d' : d :: 60 : 1.$$

Hence $G : G' :: d'^2 : d^2$, or $G : G' :: \frac{1}{d^2} : \frac{1}{d'^2}$ (233),

that is, the force of gravity at the two points is inversely as the square of the corresponding distance from the earth's centre.

From the law of gravitation, in connection with the three laws of motion, we can deduce the laws of Kepler, and also compute mathematically the amount of the perturbations of the various bodies composing the planetary system.

It should be carefully borne in mind that the attraction of the earth is not a single force, but the resultant of the separate attractions of each of the particles of which it is composed. That attraction is exerted by every particle, is shown, among other ways, by the deviation of a plumb-line from the vertical when near the side of a mountain. The irregularities on the earth's surface are, however, so small in proportion to the magnitude of the whole globe, that they may generally be neglected in astronomical calculations.

320. Attraction upon a Body situated on the Surface of a Sphere. The principles stated in the preceding paragraphs, furnish a method of ascertaining the relative attraction of two spheres upon a body of given mass situated at their surface. Let A, B, be two spheres of equal density, with radii R, R', respectively. Since the total mass of each may be imagined to be concentrated at its centre, calling G, G', their attractions, M, M', their masses, we have, $G : G' :: \frac{M}{R^2} : \frac{M'}{R'^2}$. But since the masses of the spheres are proportional to the cubes of their radii, $M : M' :: R^3 : R'^3$. Hence

$$G : G' :: \frac{R^3}{R^2} : \frac{R'^3}{R'^2}, \text{ or } G : G' :: R : R' \quad (234).$$

That is, *the attractions of spheres of equal density on a body situated upon their surface are proportional to the radii of the spheres.*

If the densities vary, the masses of the spheres are proportional to the volume multiplied by the density, that is, $M : M' :: Vd : V'd' :: R^3d : R'^3d'$. Whence

$$G : G' :: \frac{R^3d}{R^2} : \frac{R'^3d'}{R'^2} :: Rd : R'd'. \quad (235).$$

To show the application of this proposition, suppose that it is required to find the relative weight of a body at the surface of the earth, and at the surface of the sun. The mean radius of the earth is, in round numbers, 6377 km., that of the sun 709,700 km. The density of the sun is but one-fourth that of the earth. Hence from (235) denoting by G, the weight of a body at the earth's surface, and by G', its weight at the surface of the sun, $G : G' :: 6377 \times 1 : 709{,}700 \times \frac{1}{4} :: 1 : 27.8$, whence $G' = 27.8\ G$. A body weighing one kilogramme on the earth's surface would therefore weigh 27.8 kgrs. if transported to the sun.

As forces are proportional to their accelerations, a body falling to the sun, when near its surface, would acquire an acceleration of 9.8087×27.8 m., or 272.68 m. per second.

321. Diminution of Gravity above Surface of Sphere. Since the resultant effect of the attraction of all the particles of matter composing a sphere upon a body outside of it, is the same as if the whole mass were concentrated at the centre of the sphere, it follows that the attraction exerted upon any body varies inversely as the square of its distance from that centre.

Hence if we call G, G', the attraction upon the body at the surface of the sphere, and at a distance h from the surface, $G : G' :: \frac{1}{R^2} : \frac{1}{(R+h)^2}$ (235), whence $G' = G\frac{R^2}{(R+h)^2}$ (236), or performing the division, we have approximately $G' = G\left(1 - \frac{2h}{R}\right)$ (237).

This formula can evidently be used to determine the weight of a body when elevated above the surface of the earth. Thus a body weighing 1 kgr. at the surface, if carried in a balloon to the height of 5 km. above the surface would have for its weight $G' = 1(1 - \frac{10}{6377}) = .998$ kgrs.

The acceleration g', which a falling body would acquire at any given elevation, may also be found from proportion (235). For $G : G' :: g : g' :: \frac{1}{R^2} : \frac{1}{(R+h)^2}$, whence $g' = g\frac{R^2}{(R+h)^2}$ (238); approximately, $g' = g\left(1 - \frac{2h}{R}\right)$ (239). Also $g = g'\left(1 + \frac{2h}{R}\right)$ (240) approximately.

322. Gravity within Hollow Sphere. *A body placed anywhere in the interior of a spherical shell of uniform thickness and density, will be equally attracted in all directions.*

Let P be a body placed at any point within the spherical shell $ABCab$, and subject only to the attraction of the matter composing that shell, and let ab be an element of the surface. The lines of attraction of the particles composing ab will be comprised within a cone aPb, whose base is the element ab, and whose vertex is P and their resultant will be in the axis of the cone. The elements of Pab, if prolonged through P, will form another cone, PAB, having a vertex P and base AB. The lines of attraction of AB will evidently lie within the cone PAB, and their resultant will coincide with its axis, being therefore directly opposed to the resultant of the attraction of ab. Hence

$$\textit{Attraction of } ab : \textit{Attraction of } AB :: \frac{ab}{Pa^2} : \frac{AB}{PA^2}.$$

But $ab : AB :: Pa^2 : PA^2$.

Hence

$$\text{Attraction of } ab : \text{Attraction of } AB :: \frac{Pa^2}{Pa^2} : \frac{PA^2}{PA^2} :: 1 : 1. \quad (241.)$$

The body P is therefore equally attracted towards ab and towards AB. A like demonstration can be applied to every other element, and to any number of concentric shells; hence the body is in equilibrium.

323. Gravity below Surface of Sphere. Suppose a body to be situated at any point B, below the surface of a sphere AFE, Fig. 174. Denote AC by R, BC by R'. The resultant effect of that portion of the sphere outside of BHD is zero, by the preceding proposition. The whole resultant attraction at B is therefore that of the sphere BHD of radius R'. But the attraction when at A is that of the sphere AFE, hence calling G, G', the attractions at A and B, respectively, we have from (235), $G : G' :: Rd : R'd'$, or as the densities, d, d', are the same, $G : G' :: R : R'$. (242.) Hence *the gravity of a body situated within a solid sphere is directly proportional to its distance from the centre.*

This proposition can not be applied to determining the weight of a body below the surface of the earth, as the density varies with the depth. The law in this case will be explained in treating of the *mass of the earth.*

324. Effect of Spheroidal Form of Earth on Gravity. Since the earth is not perfectly spherical, but flattened at the poles, it is evident that its attraction upon a body will not be the same at every portion of the surface. Analysis shows that gravity increases from the equator to the poles. The loss of weight, due to the want of sphericity of the earth, in the case of a body carried from either pole to the equator is $\frac{1}{568}$ part of the total amount.

325. Effect of Centrifugal Force. The actual change of weight, as computed from the results of experiments, is much greater, being $\frac{1}{191}$ part of the whole weight. This is owing to the fact that besides the diminution explained in the preceding paragraph, there is another and greater cause of decrease, owing to the centrifugal force due to the earth's rotation, one component of which is always opposed to the weight of terrestrial objects, and which increases as the latitude decreases.

An approximate formula, showing the effect of the centrifugal force on the weight can be demonstrated from Fig. 175. Let the centrifugal force at R, the equator, be denoted by F, that at E, in latitude θ, by F'. Then $F' = F \cos \theta$, as shown in § 283, p. 137. Let EG represent the centrifugal force F' at E. Resolving this into two components, one EH, perpendicular to the surface of the sphere, the other EK, tangential to it, $EH =$

$EG \cos GEH = EG \cos \theta$, whence if the component represented by EH be called F'', $F'' = F' \cos \theta = F \cos^2 \theta$ (243), which represents the loss of weight due to the centrifugal force when F is the loss due to that force at the equator. This demonstration evidently supposes the earth to be spherical.

326. Corrected value of g. It can be shown that if g be the acceleration due to gravity in latitude 45°, and at the level of the sea, the acceleration, g', in any other latitude, L, at the level of the sea, will be $g' = g\,(1 - 0.002552 \cos 2L)$. (244.) For any elevation, h, above the sea-level, the formula becomes $g' = g\,(1 - 0.002552 \cos 2L)\left(1 - \frac{2h}{R}\right)$. (245.) This value of g' is that which would occur in mid-air, as in the case of a body let fall from a balloon. In the case of mountain summits, or table-lands, a somewhat different formula is generally used.[1] The existence of local disturbances renders it im ossible to derive any general formula which will give the exact value of g for any particular place. Hence this can only be determined by direct observation by methods to be detailed shortly.

327. Historical Sketch. The way to the discovery of the law of gravitation was opened by the establishment of the fact that the sun is the centre of the planetary system, as taught by Copernicus,[2] whose work, *De Revolutionibus Orbium Cœlestium*, was published in the year of his death, 1543, although he had promulgated his doctrines more or less extensively some years before. During the latter part of the 16th, and early part of the 17th centuries, Kepler[3] discovered the three laws bearing his *name*. The first and second laws were made public in his work, *On the Motions of Mars*, published in 1609, the third in the *Harmonice Mundi*, published in 1619, though the third law was surmised long before either of the others. Shortly after, Galileo[4] discovered the laws of momentum, and by observations with the newly invented telescope, added fresh confirmation to the truth of the Copernican System.

From the three laws of Kepler, in connection with the laws of momentum, Newton[5] established the law of universal gravitation, as already described, during the years between 1666 and 1687. The various facts established by Newton in completing his theory are as follows:[6]—

1. The force by which the *different* planets are attracted to the sun is in the inverse proportion to the squares of their distances.

This was shown to result from the third Law of Kepler.

2. The force by which the *same* planet is attracted to the sun in different parts of its orbit, is also in the inverse proportion to the square of the distances.

This proposition follows from the first and second laws of Kepler, as shown in § 273.

3. The *earth* also exerts such a force on the moon, and this force is identical with the force of *gravity*. (See § 319.)

[1] See Chapter on *Pendulum*.

[2] Born at Thorn, Prussia, in 1473; died, 1543.

[3] Born at Magstatt, Würtemberg, 1571; died, 1630.

[4] Born at Pisa, in 1564; died, 1642.

[5] Born at Woolsthorpe, 1642; died, 1727.

[6] Whewell's *History of the Inductive Sciences*, Vol. I, p. 399. (Appleton's Edition, New York, 1865.)

4. Bodies act thus on *other* bodies, besides those which revolve around them; thus, the sun exerts such a force on the moon and satellites, and the planets exert such forces on *one another*.

5. The force thus exerted by the general masses of the sun, earth and planets, arises from the attraction of *each particle* of these masses; which attraction follows the above law, and belongs to all matter alike.

The last two propositions were demonstrated from various astronomical phenomena by the most acute mathematical analysis, and Newton frequently found it necessary to invent new methods for the solution of the particular problems which constantly arose in the course of his research.

The theory of gravitation has been greatly extended since the time of its originator, so as to explain quantitatively the various planetary perturbations, and other problems of astronomy, by the labors of Laplace, Lagrange, D'Alembert, Clairaut, and many other scarcely less celebrated scientists.

References.

Philosophiæ Naturalis Principia Mathematica, by Sir Isaac Newton.

Treatise on Infinitesimal Calculus, by Bartholomew Price, Vol. III.

Treatise on Analytical Statics, by Isaac Todhunter. (London: Macmillan & Co.)

Mathematical Principles of Mechanical Philosophy, by John Henry Pratt. (Cambridge, 1836.)

History of the Inductive Sciences, by Wm. Whewell, Books V., VI., VII.

History of Physical Astronomy, by Robert Grant; (London: Henry G. Bohn.) Chapters I – XIII.

Laws of Falling Bodies.

328. Laws. We now pass to the consideration of the laws governing the fall of bodies through the air. We shall, for the present, consider the distance fallen through to be so small that the force of gravity throughout that distance may be considered as uniform.

1. *The velocity of a falling body is independent of its mass, whence all bodies fall with the same velocity.*

Since gravity is directly proportional to the mass, if G, G', be the forces acting on two bodies whose masses are M, M', respectively, we have, $G : G' :: M : M'$. Also, the momenta generated by these forces in equal times, are proportional to the forces themselves; hence, calling V, V', the velocities acquired by M, M', in equal times, $G : G' :: MV : M'V'$, whence $M : M' :: MV : M'V'$, and $V = V'$.

This law was unknown until the time of Galileo, earlier philosophers having supposed that bodies fell with velocities proportional to their masses. Galileo, however, reasoned that the aggregate force acting upon a number of equal separate falling bodies, a collection of balls, for instance, would neither be increased nor diminished by uniting them in a single mass, so that each particle of such a body would fall with the same rapidity as if it were free. He verified his reasoning in an experiment performed about

1590, in which he dropped simultaneously balls of different material and weight from the summit of the Leaning Tower of Pisa, the balls being found to strike the ground at the same instant.

2. *The velocity is independent of the material of which the falling body is composed.*

This follows directly from the fact that gravitation depends simply upon the mass and distance of bodies, and in no way on their composition.

Experience seems in a measure to contradict these two laws. Thus, a ball of metal falls more rapidly than a sheet of paper. This arises, however, from the resistance of the air, which buoys up the extended surface of the paper, while it offers comparatively little opposition to the passage of the ball. If this disturbing force be removed, the two fall with equal velocities. This is illustrated by an experiment originated by Newton. A long tube from which the air can be exhausted, contains a coin and a feather. On holding the tube vertically these are seen to fall, the coin very rapidly, the feather very slowly. If the air is now removed from the tube by means of an air-pump, and the experiment repeated, the two will fall equally fast.

3. *The velocity is proportional to the time of descent.*

4. *The velocity is proportional to the square root of the distance fallen through.*

5. *The spaces traversed by a falling body are proportional to the squares of the times occupied in describing them.*

The truth of the last three laws may be inferred from the fact that falling bodies, being acted on by gravity, a constant force, their motion must be uniformly accelerated.

We have next to explain the means by which these laws are verified.

329. Inclined Plane. The simplest method of demonstration is the one originally used by Galileo, and explained by him in his *Dialogues on Motion.* Since bodies rolling down an inclined plane are acted upon by a constant accelerating force, their motion is uniformly accelerated, and as $a = g\frac{H}{L}$ (p. 125), the less the height in proportion to the length, the less will be the acceleration, so that by giving a very small inclination to the plane, the body may be made to roll with sufficient slowness to measure the spaces traversed in successive seconds, which is difficult in the case of a body falling freely. It is merely necessary to fix a scale of equal parts upon the plane (Fig. 176), and letting a ball start from its summit, to note the number of divisions that it passes over in 1, 2, 3, etc., seconds. They are found to bear to each other the relation of 1, 4, 9, etc., the squares of the times, which involves all the laws of uniformly accelerated motion.

As $s = \frac{1}{2}gt^2\frac{H}{L}$ (p. 126), we can substitute for s, t, H, L, their values, as thus measured, and by solving the equation relatively to g, obtain the value of the acceleration of gravity. The friction of the ball, however, renders the value as thus determined, somewhat too small.

330. Attwood's Machine. Another method of verifying these laws is by the apparatus known as *Attwood's Machine*, from the name of the inventor. Its principle is very simple. Suppose two equal weights, P, P, to be suspended from the opposite ends of a thread passing over a pulley A, Fig. 177. These will remain in equilibrium so long as undisturbed, but if a small additional weight, p, be placed upon one of them, it will descend with an acceleration dependent upon the magnitude of the weights P, P, p. To find this acceleration we must observe that were p to fall freely, the velocity generated in one second would be g. But in the apparatus the fall of p puts the weights P, P, and the pulley A in motion, as well as the mass of p itself. Hence as the the momenta generated in one second must be equal in the two cases, if we neglect the effect of the weight of the pulley A, we have from proportion (5), p. 34, $a : g :: p : 2P + p$, whence $a = g\frac{p}{2P + p}$ (246). If, therefore, we make p very small, as compared with $2P$, the motion will be so slow that the spaces passed over in successive seconds can be read on a scale placed behind the falling weight.

In the preceding demonstration we have, for simplicity, neglected the effect of the mass of the pulley A, and of the cord connecting P, P. This is not generally allowable in practice, as the pulley is usually quite large. We must therefore modify equation (246) so as to take this into account. The correction to be applied is due to the inertia of the pulley and cord, which must be overcome in order that A may revolve with the same velocity as that possessed by the weight. This resistance is a constant, and produces the same effect as if there were an additional weight P', to be moved by the gravity of p. Modifying (246) in accordance with this fact we have $a = g\frac{p}{2P + P' + p}$. (247.) The constant P', is best found by direct experiment. A known weight, p, is placed on P, and the acceleration a is measured. The values thus obtained are substituted in equation (247), in which P' remains the only unknown quantity, which is therefore immediately determined once for all.

The most approved form of Attwood's machine is represented in Fig. 178. The pulley A is mounted on friction wheels B, B, in order to reduce the resistance caused by friction. The time of descent is measured by a second's pendulum C, which is so ar anged as to *make* and *break* an electric circuit at every vibration, thereby causing an electro-magnet E to act upon a bell-hammer, so that the bell sounds once every second. The corresponding spaces are read upon the graduated scale LM. T is a table capable of sliding along this scale, and upon which the descending weight is received.

To verify the law of the proportionality of the spaces to the square of the times, the weight P, together with a small additional weight p is placed upon a little table M. This table is connected with the armature of the electro-magnet E in such a manner that when the circuit is completed by the pendulum, it is detached, thus allowing the weight to fall with a uniformly accelerated motion. The weight being in place, the pendulum is made to vibrate, and simultaneously with the stroke of the bell, the table is detached, and the weight begins to fall. By repeated trial the table T is then adjusted so that the click of the descending weight as it strikes, and the sound of the bell striking the second second, shall coincide. The corresponding length of the scale is noted, and the experiment is then repeated, using intervals of 1, 2, 3, etc., seconds. The lengths of the corresponding spaces on LM are then found to be proportional to 1, 4, 9, etc., the squares of the times of descent.

The law of the proportionality of the velocity to the time can also be proved directly with this apparatus. For this purpose there is a second table, M, consisting of a ring of metal. The weight p is made in the form of a bar, so that it shall be lifted from P by this ring, leaving the larger weights P,P, to move onwards uniformly, with the velocity possessed by them at the moment of the withdrawal of the accelerating force. If the table B is placed so that the weight is removed at the end of 1, 2, 3, etc., seconds, and the space passed over during the succeeding second is measured, these will be the velocities generated in 1, 2, 3, etc., seconds, which will be found to bear to each other the relation of the numbers 1, 2, 3, etc., that is, the velocities are proportional to the times occupied generating them.

By determining a by experiment, and solving equation (247) relatively to g, the value of the acceleration due to gravity may be found.

331. Barbouze's Machine.[1] A modification of Attwood's Machine has been constructed by Barbouze of Paris, which is designed to give more accurate results than it is possible to obtain with the former apparatus. The axis of the pulley of an Attwood's Machine is attached to a light, cylindrical drum, covered with lampblacked paper. Against the paper rests a style attached to a tuning-fork. A small additional weight is added at one end of the cord, as already described, and after having set the fork in vibration, the weight is allowed to fall. The drum, moving with the pulley, assumes a uniformly accelerated motion, and the vibrating style removes the lampblack whenever it touches, leaving a sinuous line, the undulations of which are described in equal times, as the vibrations of the fork are isochronous. Owing to the nature of the motion of the drum, the distance between corresponding points of the sinuosities increases, the distance of each from the beginning of the curve measuring the space described by the

[1] See *Elementary Treatise on Natural Philosophy*, by A. Privat Deschanel; translated by J. D. Everett; Part I., p. 45. Also *Report of U. S. Commrs. to Paris Universal Exposition*, Vol. III, p. 489.

revolving cylinder. These distances are found to bear to each other the relation 1, 4, 9, 16, etc., which are proportional to the squares of the times from the beginning of the motion.

In the machine, as actually constructed, the vibration of the tuning-fork is frequently kept up by electricity, and the weight is allowed to fall by breaking an electric circuit. The best way of marking the intervals of time is first to allow the instrument to revolve while the fork is not in vibration. The style then traces a line without undulations, *ab*, Fig. 179. If the fork is now sounded, and the body allowed to fall, a sinuous line, *acdb*, is described, the portions *ac*, *cd*, *db*, lying between its intersections with *ab*, being drawn in equal intervals of time. The other laws of accelerated motion can evidently be verified by removing the weight when part of the descent has been accomplished, as described in the case of Attwood's Machine, after which the spaces *ac*, *cd*, *db*, will be equal to each other.

A similar method has been used in the laboratory of the Institute, in which a freely-falling glass plate covered with lampblack, is pressed against a style attached to a vibrating fork.

332. Horizontal Accelerating Machine. This apparatus is similar in principle to Attwood's Machine. A wagon is arranged so as to move over a horizontal table, *AB*, Fig. 180, the motive force being the small weight, *p*, attached to it by a cord running over a pulley. A scale is placed on the table so that the position of the wagon can be noted at any moment by means of an index *I*. To demonstrate the law of the spaces, it is merely necessary to observe the number indicated by the index, at the expiration of 1, 2, 3, etc., seconds from the beginning of the motion. If P be the weight of the wagon, cord, etc., p the added weight, the acceleration $a = g\frac{p}{P + p}$. (248.)

333. Morin's Machine. In this apparatus, which is represented in Fig. 181, the falling body P is allowed to descend freely under the influence of gravity, being guided in its descent by two vertical wires, *KL*, *IS*. *MN* is a cylindrical drum (in the original apparatus about 2 metres high), which is covered with paper, and is made to revolve uniformly about a vertical axis by means of clock-work. A pencil attached to P presses against this paper, so as to make a mark when the body falls. Evidently if the cylinder were at rest the line would be vertical; while if the body remained at rest, and the cylinder revolved, the line traced would be a horizontal circle, *OX*. But if both *MN* and P are in motion simultaneously, a curve will be traced, the nature of which depends on the law of motion of the falling body. If the motion of the body is uniformly accelerated, it is plain that calling E, F, two points on the curve, described in t and t' seconds, respectively, the vertical lines *AE*, *BF*, which are the corresponding spaces described by the falling body, will be proportional to the squares of the horizontal lines OA, OB, which measure the times, since the motion of the cylinder is uniform. That is, if the paper be taken from the cylinder and spread upon a plane surface, the curve will be similar to that shown in Fig. 182, in which, for two points, E, F, whose coördinates are (xy), $(x'y')$, we have, $x^2 : x'^2 : : y : y'$. The curve possessing this characteristic is a *parabola*. Now the curve actually traced by the falling body is found to be a parabola, hence its motion must be uniformly accelerated.

334. Additional Methods. By far the best method of verifying the laws of falling bodies is by the use of some of the forms of *chronograph*, in which the registering of the time of descent is accomplished by electricity. The delicacy of such instruments far transcends any of those which we have described, but as the understanding of the apparatus involves a knowledge of the principles of electricity, they will be treated under the practical applications of that agent.

335. Case of Body falling from great Height. In treating of the laws of falling bodies we have supposed them to be acted upon by a uniformly accelerating force. This is not strictly true, however, for gravity varies inversely as the square of the distance, so that for a body falling from a very great elevation, the acceleration would increase as the body approached the earth. But in most problems in Physics, the height fallen through is comparatively small, so that the diminution of gravity above the surface may be neglected. Indeed, at the height of a kilometre it would be but $\frac{1}{3188}$ of the total weight of the body.

As the force acting upon a falling body varies inversely as the square of the distance from the centre of the earth, it can be shown by the calculus that there is a limit to the velocity which can be possessed by a body falling to the earth. It can be proved that a body falling from an infinite distance would acquire a velocity of about 11 kilometres per second.

Projectiles.

336. Statement of Problem. In treating of the theory of projectiles we assume that the moving body is acted upon by but two forces; (1) the original impelling force, and (2) the force of gravitation. The disturbing effects caused by the resistance of the air are computed separately, and applied as corrections.

Since the greatest horizontal distance to which a projectile can be thrown is comparatively very small, amounting to but a few minutes of arc, the directions of the force of gravity at all points of the path may be supposed to be parallel. It may also be regarded as invariable in intensity, since the greatest vertical height ever reached is less than a mile.

337. Path of Projectile. To ascertain the path pursued by a body thrown into the air, let AB, Fig. 183, represent the direction in which it is projected. Were there no deflecting or resisting force, the body would move uniformly in this line, but since gravity is constantly drawing it towards the earth, it will, at the same time, have impressed upon it a uniformly accelerated downward motion. The resultant of these movements will be the path sought. Hence the point occupied by the projectile at the end of any interval of time, can be found in the following manner. Let v be the velocity of projection. Then at the end of 1 second the body will have moved over a distance $AG = v$, by virtue of the projectile force, while under the action of gravity it has fallen through a vertical distance $GC = \frac{1}{2}g$. The body will therefore be found at C, at the expiration of the first second. In like manner, at the end of 2 seconds the position of the body will be found by laying off $AH = 2v$, and from this point H, letting fall a vertical $HD = 2g$. At the end of 3 seconds it will be at E, AI being equal to $3v$, and IE to $\frac{9}{2}g$. So after 4 seconds it will be at F, and after t seconds at M. The path of the projectile is therefore a curve passing through these points. As $AG = v$, $AB = vt$, $GC = \frac{1}{2}g$, $BM = \frac{1}{2}gt^2$, $GC : BM :: \frac{1}{2}g : \frac{1}{2}gt^2 :: 1 : t^2$. Also $AG : AB :: v : vt :: 1 : t$, and $AG^2 : AB^2 :: 1 : t^2$, whence $GC : BM :: AG^2 : AB^2$. (249.) The curve is therefore of such a nature that GC varies as AG^2, or the deflection is proportional to the square of the corresponding tangent, a property characteristic of the *parabola.*

The time occupied in describing the curve $AEMN$, is called the *time of flight*; the horizontal distance AN from the point of starting to the point at which the projectile reaches the earth, is the *horizontal range*; and the greatest elevation attained by the body is the *vertical range*, or *flight.*

These quantities are dependent upon the initial velocity of the projectile and the angle of projection, and when the latter are given, can easily be found. We confine our investigations to the case in which the line AN is horizontal.

338. General Equations. Let AG, Fig. 184, be the line of projection, making an angle, θ, with a horizontal plane, AN. Call V the velocity of projection, which may be resolved into two components, one of which, $V_h = V \cos \theta$ (250), is horizontal, the other, $V_v = V \sin \theta$ (251), vertical. During the movement of the projectile the horizontal component, V_h, is constant, hence the horizontal distance passed over in time t will be $S_h = V_h t$. The vertical component, V_v, is, however, uniformly diminished by gravity, according to the laws of uniformly retarded motion; hence, after t seconds, the vertical component will have become $v = V_v - gt$, and the space passed over in that time will be,

$S_v = V_v t - \frac{1}{2}gt^2$. We have, therefore, the following fundamental equations:—

$v_h = V_h = V \cos\theta$ (252), $S_h = V_h t = V \cos\theta t$ (253), $v_v = V_v - gt = V \sin\theta - gt$ (254), $S_v = Vt - \frac{1}{2}gt^2 = V \sin\theta t - \frac{1}{2}gt^2$ (255).

Equations (252), (254), give the vertical and horizontal components when the initial velocity V, the angle of elevation θ, and the time from the beginning of the motion is known; equations (253), (255), the coördinates of the projectile after the lapse of any given number of seconds.

339. **Time of Flight.** In ascending from A to M, the highest point reached by the projectile, that is, in half the time of flight (§ 249, p. 122), the total vertical component of the motion is destroyed by gravity. Hence, calling T the time of flight, we have $\frac{gT}{2} = V_v$, or $gT = 2V \sin\theta$, whence $T = \frac{2V \sin\theta}{g}$ (256), a formula giving the time of flight in terms of V and θ.

340. **Vertical Range.** Since the time of descent equals half the time of flight, or $\frac{T}{2}$, the vertical range, R_v, must be $\frac{1}{2}g\left(\frac{T}{2}\right)^2$, or substituting the value of the time of flight as given in (256), $R_v = \frac{V^2 \sin^2\theta}{2g}$. (257.) This value may also be obtained by substituting for t, in (255), $\frac{T}{2}$, and inserting the value of T given in (256).

341. **Horizontal Range.** From (253), $S_h = V \cos\theta t$, in which if $t = T$, we have the horizontal range,

$$R_h = V \cos\theta T. \quad (258.)$$

Also since $T = \frac{2V \sin\theta}{g}$, $R_h = \frac{2V^2 \sin\theta \cos\theta}{g} = \frac{V^2 \sin 2\theta}{g}$ (259), a formula giving the horizontal range in terms of V and θ.

342. **Relation of Range to Angle.** In equation (259), R_h varies as $\sin 2\theta$, and is therefore a maximum when $\sin 2\theta = 1$, in which case $\theta = 45°$. Hence with a given velocity, the greatest horizontal range is obtained when the angle of elevation is 45°. Also since $\sin 2(45° + \phi) = \sin 2(45° - \phi)$, ϕ being any angle, the horizontal range is the same for angles equally above or below 45°. The maximum vertical range evidently occurs when $\theta = 90°$, and equals $\frac{V^2}{2g}$. (257). Since $R_h = \frac{V^2 \sin 2\theta}{g}$, if $\theta = 45°$, $R_h = \frac{V^2}{g}$. But $\frac{V^2}{2g}$ is the maximum vertical range, hence with a given velocity the greatest horizontal range is twice the greatest vertical range.

If $\epsilon = 0°$, and the projectile is fired from an elevation, the time of flight is equal to the time in which it would fall vertically from that elevation to the earth. This follows from (255), in which, by substituting T, the time of flight for t, S_v becomes equal to R_v, and $R_v = -\frac{1}{2}gT^2$, the space which would be traversed in T seconds by a body falling freely.

343. Equation of Trajectory. To those familiar with the processes of analytical geometry, the following demonstration may be more satisfactory than the preceding. The equation of the curve OBM, Fig. 185, in which the projectile moves, is $y = x \text{ tang } \theta - \frac{g}{2V^2 \cos^2 \theta}x^2$ (260), which is the equation of a parabola with a vertical axis. The equation is derived in the following manner. Let B be any point on the parabola, with coördinates x, y, reached by the projectile in t seconds, after the beginning of its motion. Then $y = CA - AB = x \text{ tang } \theta - AB$. But $AB = \frac{1}{2}gt^2$, and as $x = V \cos \theta t$, $t = \frac{x}{V \cos \theta}$, whence $AB = \frac{gx^2}{2V^2 \cos^2 \theta}$, whence $y = x \text{ tang } \theta - \frac{g}{2V^2 \cos^2 \theta}x^2$. To find the horizontal range, in (260), make $y = 0$, and solve relatively to x, which will be found to be equal to $\frac{2V^2 \cos^2 \theta \text{ tang } \theta}{g} = \frac{V^2 \sin 2\theta}{g}$, as before. To find the vertical range, the maximum value of y should be found from equation (260), regarding x and y as variables.

344. Parabola of Safety. The curve tangent to all the parabolas, ACF, ADH, ANE, etc. Fig. 186, in which a projectile moves when the angle of elevation varies from 0° to 90°, is a parabola, and is known as the *parabola of safety*, because no point beyond it can be reached by a projectile fired with a given velocity. The range of a projectile is therefore included within a paraboloid of revolution, found by rotating the parabola of safety about the vertical axis AB. In Fig. 186, $ACDE$ is the parabola of safety in the case of a projectile whose maximum horizontal range is AE.

345. Effect of Resistance of Air. The theoretical curve of motion of any projectile being ascertained, it remains to see how this is modified by the resistance of the air. A moving body in passing through the atmosphere pushes the gaseous particles aside, thus losing a certain amount of energy, which is evidently a measure of the resistance offered.

The resistance is evidently proportional to the density of the medium, and to the surface of the projectile, since the mass of the air displaced varies directly as these quantities. Supposing no farther action to be exerted by the displaced particles, the resistance also varies as the square of the velocity of the projectile. For suppose two balls A and B, of equal masses, A moving with x times the velocity of B. A communicates x times the velocity given by B, to each particle it displaces, and displaces x times as many particles in a given time, thus experiencing x^2 times the resistance.

But there is another consideration. Only in the case of very small velocities is the supposition allowable that the air has no further effect after coming in contact with the projectile. When the velocity is considerable, the air in front of the body becomes compressed, that in front rarefied, thus greatly increasing the resistance.[1]

Recent experiments of Professor Bashforth have shown that for velocities from 1100 to 1400 feet per second, the total resistance to motion varies approximately as the cube of the velocity, which is a sufficiently accurate ratio for practical purposes. At velocities of from 1400 to 1700 ft. the resistance is nearly as the square of the velocity.

The actual resistance in pounds upon a spherical shot, and upon an elongated projectile with an ogival head is shown in the following table, abridged from the results of Bashforth.[2]

Resistance in Pounds.

Velocity in feet.	*Spherical Shot. Diameter.*				*Elongated Shot. Diameter.*			
	1 *in.*	5 *in.*	10 *in.*	15 *in.*	1 *in.*	5 *in.*	10 *in.*	15 *in.*
1000	4.4	110	438	986	2.3	58	233	524
1250	9.2	229	917	2063	6.6	165	660	1484
1500	14.1	351	1406	3163	10.2	255	1019	2293
1750	19.5	439	1954	4397				
2000	25.8	645	2582	5809				

The effect of this resistance is of course to cause an enormous diminution of range, as shown in the following table, which gives the range in yards of a 32 lb. shot, fired with an initial velocity of 1600 ft., in vacuo and in air.

Elevation.	1°.	2°.	3°.	4°.
Range in Vacuo.	930	1840	2786	3709
Range in Air.	780	1160	1460	1690
Ratio.	1.19	1.58	1.90	2.19

Elongated shot are retarded less than balls, because they are better fitted for cleaving the air.[3] The long ranges of rifled guns are largely due to the use of elongated shot, the initial velocity being much less than in the case of smooth-bore arms.[4] An *ogival head*, Fig. 187, offers the least resistance.

1 See Hutton's *Tracts, Mathematical and Philosophical*, No. 37.

2 See Owen's *Modern Artillery*, p. 434.

3 See preceding table.

4 Owen's *Modern Artillery*, p. 12.

346. Effect of Resistance on Trajectory. The rapid destruction of moving force by the resistance of the air causes the trajectory of a projectile to vary greatly from a parabola for all velocities over 200 to 300 ft. per second, the latter portion of the path being much more inclined to the horizon than theory would indicate. Also, for the same reason, the practical angle of maximum range is not 45°, except for very low velocities. For rapid projectiles this angle is 35°.

347. Effect of Rotation of Projectile. If the ball moves onward with no rotary motion, it is clear that the resistance of the air will retard all parts equally, and hence there will be no tendency to deviate in any direction. The same is true if the ball rotates about the line of flight as an axis, as in Fig. 188. If, however, there is a rotation about an axis at right angles to the line of flight, Fig. 189, it is evident that there will be a greater resistance at *A* than at *B*, because at the former point the projectile and rotary motions are in the same direction, while in the latter they are opposite. The resistance at *A* is therefore due to the sum of these velocities, that at *B* to their difference.

Experiments by Magnus[1] show that in such a case there would be a greater atmospheric pressure at *A* than at *B*, hence the projectile must deviate towards *F*.[2] Were the rotation in the opposite direction, the direction of the deviation would be reversed. Hence any ball thus moving will be deflected to the right or the left, or its range will be varied, according to the direction of its rotation.

Such a motion is invariably acquired by a ball fired from a smooth-bore gun, because the bore being necessarily a little larger than the diameter of the projectile, this rebounds from one point to another in its passage to the muzzle, Fig 190. Hence a rotary motion is imparted, the direction of which depends chiefly on the last point of impact. Thus a ball striking at *B*, just before leaving the gun, proceeds with a right-handed rotation.

This source of deviation can be avoided by *rifling* the gun. By means of spiral grooves cut in the bore, the shot is made to assume a rotary motion about the axis of the piece, so that on starting this rotation is at right-angles to a tangent to the trajectory. The pitch of the screw varies in different guns, but the projectile generally makes less than one complete revolution before leaving the piece.

The rotation caused by rifling also neutralizes the effect of any irregularities of surface, or want of homogeneity of the projectile, which would otherwise produce a marked effect. Also an elongated projectile can be used, thus securing a much greater range. Even with rifled guns, however, there is still a deviation from the theoretical path, owing to (1) want of equality in the projectile force of the powder, (2) the effect of wind, (3) the rotation of the earth, and (4) a movement in the direction of the rotation, called *derivation*, or *drift*, caused by a combination of the forces of aërial resistance, and the rotation of the projectile, analogous to the orbital revolution of the gyroscope.

1 Owen's *Modern Artillery*, pp. 202, 221.

2 This excess of pressure is due to the fact that a layer of air in contact with the ball is set in rotation with it by friction, thus condensing the air at *A* and rarefying it at *B*.

348. Measurement of Velocity of Projectiles. It remains to describe the methods by which we determine the initial velocity of projectiles, and also find their velocity at different distances from the gun. Vari..us intruments have been devised for this purpose.

One of the earliest forms was invented by Mattei, an Italian, in 1767. It consists of a drum, AB, Fig. 191, covered with thin paper, and revolving rapidly about a vertical axis. The shot is fired so as to traverse a diameter of the drum. Were the drum at rest the two holes pierced would be diametrically opposite, but owing to the revolution, a certain arc, BE, will be described, while the shot is traversing the diameter AB, so that the point of exit of the shot will be at E instead of at B. If v is the velocity of the shot, t the time in which the drum makes a single revolution, D the diameter of the drum, and a the angle subtended by BE, $v = \frac{2\pi D}{at}$. (261.)

A later form of this apparatus, invented by Grobert, in 1804, consists of two parallel discs, A, B, Fig. 192, divided by radii into 360°, connected by a horizontal shaft, and rotating rapidly. The ball is fired parallel to the axis, and the time of its passage from one disc to the other is determined from the angle between the radii passing through the holes in each disc. The objection to these pieces of apparatus is that the time is not measured with sufficient accuracy.

A method applied by Count Rumford to obtaining the velocity of projectiles from small arms is by means of the *recoil* of the piece. Since the powder acts in one direction to impel the ball forward, and equally in the other direction to throw the gun backward, the momenta of these must be equal. Hence if we can find the velocity with which the gun recoils, knowing its weight and the weight of the projectile, we can readily compute the velocity of the latter. To find the velocity of recoil the piece is suspended horizontally by a rigid bar moving about a pivot (Fig. 193), at the end of which is an index which moves over the graduated arc BC. Calling V the velocity of the recoil, v that of the projectile, G the weight of the gun, B that of the ball, we have, $GV = Bv$, whence $v = V\frac{G}{B}$ (262). V is computed from the observed arc of vibration of the gun, being equal to $\sqrt{2gh}$, if h be taken as the versed sine of that arc.

An instrument more extensively used than any of the preceding, is the *Ballistic Pendulum* of Robins, improved by Hutton, Gregory and others. It consists of a heavy block of wood, A, Fig. 194, suspended by a rod R, moving on knife edges at P. The ball B being fired into the block of wood, causes this to swing over a certain arc denoted by the index I, which moves over the scale CD. The index is usually made to record the extent of the arc by leaving a trace on a thin layer of wax, with which CD is coated. Knowing this arc the velocity can be calculated by a somewhat complex formula founded on the laws of angular acceleration. The following is a simplification which will convey a general idea of the process. According to the laws of momentum, the momentum of the ball before striking the pendulum, and of the pendulum and ball, which after the impact move on together, are equal. Hence, calling w the weight of the ball, v its velocity, W, the weight of the block, and V its velocity, $wv = (W + w) V$, whence $v = \frac{(W + w) V}{w}$ (263). It is customary to hollow out the front portion of the pendulum, filling it with clay, so that the ball may sink into this without causing any splintering.

The methods which we have described have been used with more or less success, but more recently the invention of electric methods has given us far more accurate methods of measurement. The Bashforth, Navez-Leurs and Noble chromoscopes, in which electricity is used, have afforded most valuable results, in determining the velocity of the ball at different points of its course, both in the bore of the piece, and in its path through the air. We shall describe some of these forms when treating of electro-chronographs.

These instruments are applied to ascertain the resistance of the air by observing the decrease of velocity of the projectile as its distance from the gun increases. It is simply necessary to measure the velocity at points more and more distant from the gun. From this loss of velocity we may calculate the magnitude of the resistance by comparing it with gravity. Thus if d be the loss of velocity in t seconds, W the weight of the projectile, F the resistance in kilogrammes, observing that forces are proportional to the momenta which they destroy in equal times, we have, $F : W :: d : gt$, whence $F = \frac{Wd}{gt}$. (264.)

349. Historical Sketch. The earliest treatise on gunnery was published in 1561, by Santbach, who supposed that every projectile proceeded in a straight line till its whole velocity was spent, and then fell directly downwards. Nicolo Tartalea (1550–1554), a Venetian, represented the trajectory as first a straight line in the direction of the original projection, next an arc of a circle, pursued till its direction became vertical, and finally terminating in a vertical line reaching to the ground. He also remarked that the angle of maximum range was 45°. Rivius (1582) knew that the body continually moved in a curve, but considered the deviation at first so slight that it might be considered as a straight line. Galileo showed that the body if unresisted would move in a parabola. He also determined the law of aerial resistance for slow motions, as did Newton also. But he greatly underrated the effect of this resistance, and supposed that the actual path of a projectile was, in general, a near approximation to the curve given by theory, while it is, in reality, completely altered. Galileo's opinion, however, was accepted as conclusive, and in the works of Anderson (1674) and Blondel (1683), this theory is strenuously upheld, and tables calculated upon it as a basis. Little progress was made till, in 1742, Benjamin Robins published his *New Principles of Gunnery*. This contained the description of his newly-invented ballistic pendulum, and of his investigations on the explosive force of powder, resistance of the air and its effect in retarding projectiles. Robin's methods were still further improved by Hutton, towards the end of the last century. The general theory has not changed greatly since, but in late years the application of the electric method of registration has been of incalculable advantage in giving far more precise results than were before obtainable.

References.

Principles and Practice of Modern Artillery, by Lieut. Col. C. H. Owen. (London, John Murray, 1871.)

Treatise on Artillery, by Capt. Boxer.

United States Ordnance Manual. (Philadelphia, J. B. Lippincott & Co., 1862.)

The preceding works may be consulted for general information relative to ordnance. For the mathematical investigation of the motions of projectiles, see

Treatise on the Dynamics of a Particle, by P. G. Tait and W. J. Steele. (Macmillan & Co., 1865.) p. 76, p. 228.

Treatise on Infinitesimal Calculus, by B. Price; Vol. IV., *Dynamics of Rigid Bodies.*

Mathematical Treatise on the Motions of Projectiles, by F. Bashforth. (London, 1872.)

Traité de balistique exterieure, par N. Mayevski. (Paris, 1872.)

Tracts, Mathematical and Philosophical, by Chas. Hutton. (London, 1786.)

See also references in Owen's *Modern Artillery.*

For methods of determining velocity, see

New Principles of Gunnery, by Benj. Robins. (1742.) [*Hutton's Edition*, 1805.]

New Experiments upon Gunpowder, with an Account of a new Method for determining the Velocity of all kinds of Military Projectiles, by Sir Benjamin Thompson (Count Rumford); *Philosophical Transactions*, 1781.

Mechanics of Engineering, by Julius Weisbach; *Coxe's Translation*, Vol. I., p. 693. (*Ballistic Pendulum*).

Modern Artillery, by Lieut. Col. Owen; *Appendix* IV. *Description of Bashforth, Navez-Leurs and Noble Chronoscopes, and other methods.*

Report of Commissioners to Paris Universal Exposition, Vol. III., p. 563.

For a full description of the Noble Chronoscope for determining the velocity of the ball within the bore of the gun, see

Annales du Conservatoire des Arts et Métiers. Tome IX., No. 35. (Translated from the *Preliminary Report of the Committee on Explosives*, London.)

For historical information consult

Whewell's *History of the Inductive Sciences.* Bk. VI.

Owen's *Modern Artillery*, p. 145.

ERRATA.

P. 29, § 37, line 8. After "weight of the wood" insert "plus that of the oxygen added in the act of burning."

P. 37, § 50, line 17. For "$\frac{1}{9}$" read "$\frac{1}{10}$."

P. 43, § 59, line 2. After "towards the east," insert "its velocity relatively to the centre of the earth is 470 m. + 470 m. = 940 m. If on the other hand it be fired towards the west."

P. 46, § 65, line 6. For "Fig. 12" read "Fig. 11."

P. 49, § 76, line 10. For "pounds," read "kilogrammes."

P. 53, § 91, line 39. For "PQ" read "LQ."

P. 55, § 94, line 4. For "$A'B$" read "AB."

P. 60, § 109, line 28. For "movement" read "rotation."

P. 62, § 114. line 7. For "$P \times OH \times P + OL$" read "$P \times OH + P \times OL$."

P. 62, § 114, line 11. For "$HL = KL$" read "$HK = KL$."

P. 62, § 114, line 12. For "similarity" read "equality." For "BKL, CKH" "CGI, BGM."

P. 62, § 115, line 11. For "$F_2 B$" read "$F_2'B$."
P. 63, § 116, line 3. For "Fig. 45" read Fig. 46."
P. 63, § 118, line 6. For "$F' = 40$" read "$R = 40$."
P. 64, § 121, line 8. For "$F = F + F'$" read "$R = F + F'$."
P. 72, § 133. In equations (66), for "O" read "0."
P. 76, § 147, line 11. For "$g''' G : g'' G$" read "$g'' G : g''' G$."
P. 77, § 149, line 9. For "BF" read "BE."
P. 86, § 170, line 18. After "foot" insert "of."
P. 97, § 193, line 21. For "AD" read "AH."
P. 98, § 195, line 11. For "AP" read "AR."
P. 111, § 224. The statement relative to the angles of chisels for brass and iron, given on the authority of Lardner, is erroneous.

www.ingramcontent.com/pod-product-compliance
Lightning Source LLC
LaVergne TN
LVHW011230110826
845150LV00006B/1598

* 9 7 8 1 4 2 5 5 1 4 4 9 5 *